07.05.16
25.05.16
21~BUR

18 MAY 2017

Books should be returned or renewed by the last date above. Renew by phone **03000 41 31 31** or online *www.kent.gov.uk/libs*

CUSTOMER SERVICE EXCELLENCE

CSE

Kent County Council
kent.gov.uk

D1435981

15 Million Degrees

A Journey to the Centre of the Sun

LUCIE GREEN

VIKING
an imprint of
PENGUIN BOOKS

VIKING

UK | USA | Canada | Ireland | Australia
India | New Zealand |South Africa

Viking is part of the Penguin Random House group of companies
whose addresses can be found at global.penguinrandomhouse.com.

First published 2016
001

Copyright © Lucie Green, 2016

The moral right of the author has been asserted

Set in 12/14.5 pt Bembo Book MT Std
Typeset by Palimpsest Book Production Ltd, Falkirk, Stirlingshire

A CIP catalogue record for this book is available from the British Library

Hardback ISBN : 978–0–670–92218–5
Trade Paperback ISBN : 978–0–670–92219–2

www.greenpenguin.co.uk

MIX
Paper from
responsible sources
FSC® C018179

Penguin Random House is committed to a
sustainable future for our business, our readers
and our planet. This book is made from Forest
Stewardship Council® certified paper.

For Valerie and Alan Green,
my parents and role models

The author and publishers wish to point out that looking directly at the Sun is dangerous and can cause permanent damage to the eyes. The Sun can be viewed directly only when filters specially designed to protect the eyes are used, or if a small pin-hole projector or other indirect viewing method is employed. See the Appendix on page 271.

Contents

Introduction

This is a book about one star. Just one out of the hundreds of billions of stars that there are in our Galaxy, the Milky Way, which in turn is just one galaxy out of the hundreds of billions there are sprawled across the known Universe. When you think about it, with numbers like these, what are the chances that our star should be special? Well, after having studied our Sun for almost twenty years, it's clear to me that it is very remarkable indeed. And I want to show you why.

At the heart of the Sun a gigantic nuclear furnace provides a continual source of energy that we would love to be able to emulate here on Earth. The Sun does it naturally and will do it for around nine billion years in total. The temperature in its core is over 15 million degrees, and the material there is under immense pressure. It's in these extreme conditions that sunlight is created. I remember being startled to find out that sunlight, born as gamma radiation, takes 170,000 years to slowly trickle to the Sun's surface. There it emerges as visible light, finally pouring towards the Earth, where just over eight minutes later we are able to see it. But light isn't the only thing the Sun sends our way. And this is where things get even more intriguing.

The Sun is often a violent star and produces the most powerful explosions and eruptions in the Solar System. In this book I want to reveal to you all the sides of the Sun's character, from the serene but spotty surface we can see when we look at the Sun in visible light, to the explosive and unpredictable atmosphere that needs to be viewed through ultraviolet radiation,

X-rays and gamma rays. Both the explosions and the eruptions are beautiful to watch, but, given that we are 150 million kilometres from the Sun, it is a challenge for us to see and understand them – for that, we had to go into space.

The story of some of the early space pioneers is told in this book. We'll discover that the opportunity they had to put telescopes in space was given by redundant Second World War technology, beginning in the US in the 1940s and in the UK the decade after.

The first images of the Sun from space were taken with simple pinhole cameras. Today, the space kit we have at our fingertips includes NASA's Solar Dynamics Observatory. Telescopes on this satellite take ten highly detailed images of the Sun's atmosphere every twelve seconds. And that's just the data being gathered by one of the three instrument suites on board the satellite. This phenomenal spacecraft adds another 1.5 terabytes of data to our archives every single day. More data is generated in one week, and for one star, than the Hubble Space Telescope generates in a year for the rest of the Universe.

The space age gave us the luxury of being able to view the Sun at any wavelength in the electromagnetic spectrum and at any time, and the view above the Earth's atmosphere has been stunning. But it comes at a price. Launching 1 kilogram (equivalent to one bag of sugar) into space costs over £10,000 – giving the largest space observatories almost a billion-pound price tag. I have worked with the SOHO satellite, which was launched in 1995 and came in at a cost of £700 million. The Solar Dynamics Observatory satellite, launched in 2010, cost £550 million. But this is money that is both spent well and spent right here on Earth employing highly skilled designers and engineers across academia and industry. Our studies of the Sun are important for science but they're also important for the space sector as a whole. In the UK this industry has an annual turnover of over

£10 billion and directly employs over 34,000 people. Globally, the space economy is worth over £200 billion.

My career in science has been dominated by the space age and I want to show you just how important this period has been in transforming our understanding of the Sun. It has literally expanded our horizons and even shown that the atmosphere of the Sun reaches out to a staggering 18 billion kilometres beyond us – that's 121 times further out than the Earth's orbit. We are living in the Sun's atmosphere! And as this atmosphere changes because of the Sun's explosive side, stormy 'space weather' is felt on Earth. We'll discover how this produces a threat to modern society through its impact on our electricity distribution, satellite technology and communications. But don't worry: predicting space weather is common now and we always keep a watchful eye on the Sun for our own safety and for the sake of the space economy.

Throughout the pages of this book we'll see just how much lies behind a seemingly straightforward desire to understand the Sun – we have quite a journey ahead! We'll cover thousands of years of naked-eye observations (do not try this yourself though), hundreds of years of telescopic observations (visit a specialist supplier) and decades of observations from space (images freely available on the internet). We'll see how understanding the Sun needs the application of atomic physics, thermodynamics, electricity, magnetism, gravity and light. We'll learn how energy is transported and changed into different forms and how this discovery was made by a ship's doctor. We'll find out that the discovery of the Sun's magnetic field needed an incredible synthesis of science and engineering, but that the payoff was the birth of solar physics.

You'll probably end up thinking that I am completely obsessed by the Sun. And you'd be right. I would say that our Sun is the most important star there has ever been or ever will be, since it is the star that gave the energy required for life on

Earth to emerge and thrive. As I am writing, we are searching for life elsewhere in the Universe, both within our Solar System and beyond. We are scanning the skies for radio signals from advanced civilizations, sending out our own messages and combing the surface of Mars for microbes. And we plan to expand the locations where we are looking for life to include the moons of planets too. But, as of 2015, life on Earth, orbiting around our Sun, is all we know.

I have let other stars feature in the pages though. Their stories help us realize the full potential of the Sun in terms of the power it might one day unleash. But we'll also see that the Sun has been a stepping-stone to understanding other stars across the Universe. The extraordinary detail that we can see has provided a wealth of information. All other stars are so distant that they are seen as specks of light (apart from rare exceptions such as Betelgeuse). So it was by studying the Sun that we were able to understand that, in their most basic sense, stars are spheres of plasma that shine because of nuclear processes happening in their cores. They sound elegantly simple. And in 1926 this sentiment was put down on paper when a British mathematician, physicist and astronomer, Sir Arthur Eddington, wrote: 'it is reasonable to hope that in a not too distant future we shall be competent to understand so simple a thing as a star.'

Eddington passed away in 1944 at the age of sixty-three, just a few years before the start of the space age. He knew nothing of the Sun's million-degree atmosphere, or that we can probe the interior of the Sun using sound waves that are trapped inside it and reveal themselves by creating patches of rising and falling gas at the solar surface. If Eddington were alive today there is no question that the level and complexity of our understanding would impress him. In this book you will see just how far we have come in our quest to comprehend what he inspiringly described as something as simple as a star.

1. Light: Don't Believe Your Eyes

Some of you will be reading this book outside in the sun. Perhaps on the beach during a holiday or making the most of a sunny weekend in the garden. Or maybe you have sneaked out of work to catch up on your solar physics under the guise of a lunchtime trip to the shops. Whatever the reason, you are able to make out the curves and angles of the lines creating each letter in each word because of the sunlight falling on the pages of this book. Where the page is white, much of the sunlight is being reflected back into your eyes. Where there is ink, very little of the sunlight bounces back, and so dark shapes appear, producing the letters of the words that you are reading now.

The sunlight that you are using to read is continuously flooding from the Sun onto your page and it has travelled 150 million kilometres* to get here. Producing light is what defines our Sun as a star. All the planets, moons, asteroids and everything else in our Solar System are like the pages of this book: they are not luminous but are seen by reflected sunlight. This light that shines continuously from the Sun is an important phenomenon not only of our Solar System but also of the entire Universe. Our experience of the cosmos is mediated through the light generated by stars.

As a species we have grown up with sunlight. Every human that has ever lived has relied on it and our bodies have evolved and adapted to use it. Our evolutionary history is intimately linked to the Sun: the development of the eye that gives us sight,

* In 2012, the International Astronomical Union, the official body responsible for approving distance units in astronomy, defined the Astronomical Unit, the distance between the Sun and the Earth, to be 149,597,870,700 metres.

the synthesis of vitamin D and the pigmentation in our skin that helps protect us from sunburn and skin damage – our relationship with sunlight is essential. Not only does sunlight allow us to see, but it made our rocky planet habitable for millions of species of animals and plants. It grows our food, it drives our weather and it can even be utilized to generate electricity to power modern society. Sunlight is amazingly versatile, but above all there is something about sunlight that means it is able to keep us alive. We literally could not live without it. Sunlight's ubiquitous and central role in our lives raises the question: what is it?

The nature of light

While light is so common it can easily be taken for granted, by merely casting a shadow you know it is a . . . thing. We can block light to make a shadow, which means there is an area where light is missing. Light must be some kind of object or phenomenon, which means we have the capacity to understand it. And for generation after generation, we have been trying to work out what light is.

Despite its being an ancient question, we have really been equipped to tackle the puzzle of 'What is light?' for just the last 150 years or so. Before that, we could only investigate what light does. And there's a big difference between seeing *how* sunlight behaves and knowing *why* it does so, but one can lead to the other once the necessary scientific foundations are in place. One person in particular illuminated an aspect of what sunlight can do in 1666, when a seemingly simple experiment involving nothing more than a wedge of glass showed us why we have colour vision. The experiment was the brainchild of the legendary natural philosopher and mathematician Sir Isaac Newton.

The telescope had been invented only a few decades before Newton was born. And the primitive instruments had already been turned skyward by people such as Galileo so that their circular glass lenses could gather the light from the distant celestial object and bring it to a focus to form an image – just as the lenses in your eyes are doing now with the light reflected off the page of this book. Investigating what light is began when Newton wondered what would happen if light were passed through a piece of glass that wasn't circular in shape.

Newton darkened his room and placed a triangular wedge of glass, a prism, in a shaft of sunlight coming through a gap in the window shutters. He had expected to see a circle of colours on the wall behind, just as in a telescope. But, instead, a tiny rainbow was splashed across the opposite wall, stretching out not in a circle but in a line. In that moment, Newton discovered a whole new aspect to the behaviour of sunlight: that it can be refracted and split into its components. More importantly, he deduced that the rainbow colours are an inherent property of light itself – it wasn't the glass somehow colouring the light. Sunlight was in fact composed of a sequence of colours that when combined together looked white. He also proved it by reversing the process: a second prism could turn the rainbow back into a beam of white light again.

Sometimes after a storm we get a glimpse of what Newton observed, on a grandiose scale. When the Sun emerges from behind the clouds and its light encounters water droplets in the air, they act as prisms and the colours that make up sunlight are suddenly unravelled to produce a rainbow. Perhaps you have by your side, as you read, a glass of water. If you do, you can see this effect for yourself. Try placing the glass on a white surface in a beam of sunlight and you can glimpse the colour produced as the sunlight passes into and out of the water.

What corroborates Newton's theory is that whether you use

a prism or a glass of water or raindrops in the air, the colours of the rainbow always appear spread out in the same order: red, orange, yellow, green, blue, violet. The pages of this book only appear white because they reflect all these colours equally. We have a colourful world around us because different objects reflect differing amounts of these colours.

Newton's experiment, although illuminating, had not answered the question: what is light? Newton's guess was that light was a substance, which was made up of beams of tiny particles with the mass of the light particle varying with colour. As the light passed from the air into the glass prism and back out again, the paths of particles of different masses were bent by different amounts – refracting, or bending, the sunlight and teasing out the sequence of colours.

But Newton wasn't the only person making new observations about the behaviour of light.

Both the English polymath Robert Hooke and the Dutch natural philosopher and mathematician Christiaan Huygens were investigating the nature of light, and neither subscribed to Newton's particle theory. They had cause to doubt him as well: it had been shown that two beams of light could cross right through each other without any effect whatsoever. This is not what we would expect if light were made of particles, which could collide with each other. But it is what we would expect if light were made of waves – an idea Newton dismissed completely.

Hooke was the first Curator of Experiments at the Royal Society in London and he made contributions across many areas of science. He was also not afraid to court controversy. And when Newton published his theory of light just a few years after Hooke had published his wave theory, their scientific dispute fuelled a major rivalry between them. Newton used the successful strategy of simply outliving Hooke; he withheld the details of his complete theory of light until after Hooke's death in 1703.

In that same year Newton became president of the Royal Society. The organization continues to this very day, something I am very grateful for as I receive funding through it as a Royal Society University Research Fellow. I know the Royal Society and its building in London very well, and I've noticed that there isn't a single painting, portrait or sculpture of Hooke on display. And that's because none are known to exist, which is very odd for someone so significant within the organization. The (unsubstantiated) rumour I have heard is that when Newton became president, he disposed of everything to do with the late Robert Hooke.

Newton also outlived Huygens, who died in 1695. Huygens had developed the wave idea into a much more detailed theory, which lived on posthumously. By considering sunlight to be a wave moving through an unknown medium, similar to water waves rippling across the surface of a pond, Huygens could explain *why* sunlight behaved the way it does, matching theory to observation. Refraction (bending) is a consequence of the front surface of the wave reaching a transparent material at an oblique angle, with the part of the wave front that touches the transparent material slowing down and the rest of the wave carrying on at the same speed as before. This explains why the wave front swings around within the transparent material and changes direction. According to Huygens' theory, waves could cross each other because there was no matter to collide.

The next task in developing the wave theory of light was to work out what kind of waves they were. What was the substance that enabled light to be propagated? What exactly was 'waving'? The answer to these questions came from an unexpected direction – from the study of electricity.

Our experience of electricity is as old as our exposure to light, from lightning in a stormy sky to static electric charges on the ground. In the 1800s, over a century after we first began to

develop a fundamental understanding of light, it was electricity's
turn to be scrutinized. It was a substantial effort, and the names
of those who carried out the experimental theoretical work live
on today, in the terms for scientific units that may be familiar to
you: Ampère, Gauss, Ohm, Faraday and Coulomb. Their work
laid the foundation for a theory that united electricity with light.
The theory was developed by 1862 and was the work of the great
Scottish mathematician James Clerk Maxwell.

At the heart of electricity are electric charges, which are either
'positive' or 'negative' – arbitrary labels that have become established over the years but which reflect that there are two distinct
types. Particles with the same electric charge repel each other,
but particles with 'opposite' charges are attracted. It is this force
that is acting within every atom, keeping the negatively charged
electrons in place around the positively charged nucleus, which
contains protons and neutral neutrons, at the centre of the atom.

We all make use of moving electrically charged particles
every day: your home is full of wires carrying electrical currents
of slowly moving electrons. But something important happens
when electrically charged particles are moving, and it can be
investigated using a compass. Place a small compass (or even the
compass app on a smartphone) next to the cord of your kettle.
As you flick the electric current on and off you will see the
needle move back and forth slightly. The reason for this is that
when a charged particle moves, another force springs to life:
magnetism. And if understanding light needs an understanding
of electric currents, it must also involve magnetism.

Magnetism is another ghostly force, like electricity, which
seems to act at a distance. Electrically charged particles repel or
attract each other without ever coming into contact. In both cases
the effect can be thought of as a field of influence, visualized as a
series of field lines. Many of us are used to diagrams of magnetic
field lines from school science lessons: the field lines close in on

themselves to make continuous loops like an elastic band. If the circles are bunched close together, the field is stronger, and if they are more spread out the field is weaker. But the same applies to the less familiar electric fields. To picture the electric field of a charged particle, imagine a series of lines emanating radially out from the particle, a bit like the spokes of a bicycle wheel that stretch out from the hub. The field lines are close together near the centre, where the particle is, indicating that the force is strong there, and spread apart the farther away they stretch, showing that the greater the distance from the particle, the weaker the force. But, unlike electricity, there are no particles that have a 'magnetic' charge. Instead a magnetic field can appear whenever there is a moving electric charge. This is why one appears when you turn on your kettle and the electric current starts to flow.

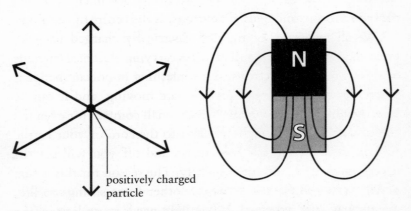

positively charged
particle

1.1 Schematics of the electric field of a stationary positively charged particle (*left*) and magnetic field lines of a bar magnet (*right*).

Don't worry if you find magnetic and electric fields some-what mysterious; it is not your fault but rather that of physics. It turns out they are two sides of the same coin, which can lead to confusing overlap.

So far we have stationary charged particles that produce an electric field and moving charged particles that produce a magnetic field (the electrons in a bar magnet are moving within their atoms, hidden from sight). But there is a different way to produce an electric field, and that is with a moving magnetic field, which is what we get when a charged particle is accelerating. But when changing its speed, the magnetic field produced by an electrically charged particle is also always in a state of flux and this in turns produces a new, secondary, electric field – an electric field which is also moving . . .

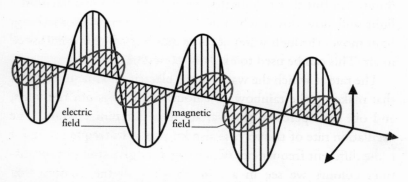

electric
field

magnetic
field

1.2 Magnetic and electric fields inducing each other in a never-ending cycle. This schematic shows that the magnetic field and electric field oscillate at right angles to each other.

Magnetic and electric fields are more intimately linked than anyone had ever expected. It was Maxwell who realized that a changing electric field, producing a changing magnetic field, producing a changing electric field, and so on ad infinitum, would produce a kind of self-propagating electromagnetic pulse that would, if left to its own devices, ripple out through space indefinitely: a self-sustaining wave not requiring any medium to move through. In 1862 Maxwell calculated the exact rate this propagation would occur at as 193,088 miles per second – almost exactly the same as the speed of light. The conclusion was unavoidable: light is an electromagnetic wave. This was the first ever meaningful answer to the question: what is light?

The conclusion that light is an electromagnetic wave solves a lot of the problems Newton was having. Waves can pass through each other without any disturbance. Electromagnetic waves also don't need a material to convey them, which is why a light wave can travel through the vacuum of space at 300,000 kilometres per second. At this speed, it takes sunlight only around 8 minutes and 20 seconds to travel the 150 million kilometres from the Sun to us. But that is only the speed it travels at in a vacuum — light will slow down when it has to pass through something: light moves through water at three quarters of its normal speed in air. This can be used to explain refraction.

The rate at which the waves in the electric and magnetic fields that make up the rainbow of colours oscillate is not the same, and our eyes perceive this variation as different colours. The oscillation rate of the wave is also known as its frequency and it is the different frequencies which together give sunlight the distinct colours we see in a rainbow (but all the colours that comprise sunlight move along at the same speed). Some light oscillates rapidly, and this is what we see as blue light, whereas red light swings back and forth much more slowly. The frequency of red light is 390 terahertz, which means 390 million million wave crests pass any given point in one second, whereas blue is up at 700 terahertz. This also means that the different colours have different wavelengths. If the waves are moving at the same speed, yet the number of wave crests passing a particular point in one second varies, the distance between the crests must vary too.

When light moves from air to water it slows down, but importantly the frequency of each wave does not change. This is compensated for by a change in wavelength. This makes no dif-

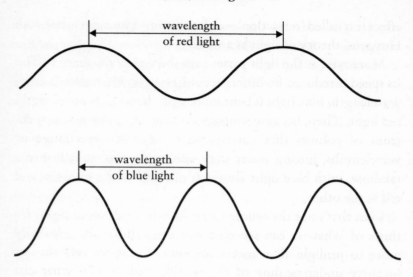

1.3 A 2D illustration of two waves with different wavelengths used to show that red light (*top*) has a longer wavelength than blue light (*bottom*).

ference to our eyes, though, because it is the frequency that they use to distinguish colour. Now, if the beam of light hits the water at exactly 90 degrees, all of the light in the beam changes speed simultaneously. But if the sunbeam hits it at an angle, different parts of the cross-section of the beam will hit the water at fractionally different times. The light that hits the water earlier will slow down before the region of light next to it in the cross-section of the beam. The result is that the light beam bends and changes its direction. Think of a group of children in line abreast and holding hands. Initially they all walk at the same speed and proceed together in a straight line. Now imagine that at one end the children walk more slowly than the others. They start to lag behind and the line of children begins to bend. They can no longer continue walking in their original direction if they keep their new speeds. When light changes direction because of this effect it is called 'refraction' – a phenomenon we met earlier with Huygens' theory of light as a wave.

Moreover, as the light passes into the water, the amount that its speed is reduced by differs according to wavelength: shorter-wavelength, blue light is bent more than the longer-wavelength, red light. Then, because sunlight consists of a continuous spectrum of colours that correspond to a continuous range of wavelengths, hitting water at certain angles will split it into a rainbow, with blue light always at one end of the rainbow and red at the other.

Even that's not the whole story. When we think of light, we think of what we can see with our eyes. But there is literally more to sunlight than meets the eye: we cannot rely on our intuitive understanding of the world, mediated by what our brains perceive, if we are going to understand the Sun. Like a detective we must be open to all possibilities, not just the ones we have been taught to expect. 'Expect the unexpected' could be the unofficial motto of modern physics.

The Sun produces light in all sorts of 'colours' we cannot even imagine; they are beyond what our visual system can perceive. We call the frequencies of light we can see 'colours', but for frequencies well outside that range we abandon the notion of colour and start giving them names like 'microwaves' and 'radio waves'. In the last seven decades of studies of the Sun we have extended our perception of sunlight into these frequencies as we have developed ways to 'see' the Sun. Telescopes and detectors on the ground and in space have given us new and different kinds of eyes. Using these we have seen that the Sun emits very low-frequency and long-wavelength light. Whereas the wavelengths of visible light are in the region of hundreds of billionths of a metre, the low-frequency waves can have wavelengths of metres. Beyond the violet light of the visible spectrum the Sun also emits high-energy ultraviolet radiation, X-rays and gamma rays that have high frequencies and wavelengths as small as atoms. This broad range of wavelengths together make what is known as the 'electromagnetic spectrum'. We need now to introduce a term that can be used to refer to any of these waves across the spectrum, whether they are visible to the eye or not. For this, the term electromagnetic radiation will be used, or radiation for short.

However, what we have found is that the Sun emits some frequencies constantly whilst others are produced in sporadic bursts. The Sun doesn't emit all frequencies of the electromagnetic spectrum in equal measure: less than half is emitted in the visible range, about half in the infrared and less than one tenth in the ultraviolet, with all other frequencies making up a small fraction of sunlight. And there's a further complication: the frequencies of radiation that arrive on Earth are not an accurate representation of what the Sun produces. The reason for this is that our atmosphere acts as a filter, and its effect is stronger on some frequencies than on others. Gamma rays, X-rays and most of the ultraviolet radiation are absorbed by our atmosphere, as is much of the infrared. Some, but not all, of the radio radiation makes its way to us. In the end, by far the biggest part of the solar spectrum that reaches us on the surface of the Earth is the Sun's visible radiation. This is a key reason why we have evolved only to detect these frequencies with our eyes: Nature is an efficient mistress and she uses the parts of sunlight that are most abundant on the ground.

There are other good reasons why Nature doesn't bother with the low- and high-frequency radiation as a means for humans to visualize the world around them. You may be wearing sunglasses as you read this book so that the intensity of the visible light reflecting off the page is reduced. When you bought the glasses I hope you chose a pair that filters out the Sun's ultraviolet radiation as well as a decent amount of the visible range: this high-frequency radiation causes eye damage and, over time, is thought to lead to cataracts.

Another excellent reason for seeing only the visible radiation becomes apparent if we consider what it would be like if our eyes were able to detect radio waves. In order to form a clear, crisp image of the world – in order to distinguish the words 'kerning' and 'keming', for example – we rely on light entering our eye through the pupil and on our eye being able to resolve the detail. Variations in the wavelength of the light and the diameter of the pupil affect the level of detail that can be achieved in this way: increasing the pupil size gives us the ability to see finer and finer detail, whereas increasing the wavelength of light has the opposite effect. An average human pupil with a diameter of, say, 3 millimetres does well for light in the visible range of the spectrum, with wavelengths between 400 billionths and 700 billionths of a metre. Consider now that the radio waves which reach us from the Sun have a wavelength of one metre – a million times greater than the wavelength of radiation used by our eyes. So, to be able to see the same level of detail as we do normally, we would need a pupil a million times larger – 3000 metres in diameter. It doesn't take eyes three kilometres big to immediately see the problem here.

Perhaps the most important thing that we can infer from sunlight is the simplest thing of all. And we can observe this without using our eyes at all. If you are reading this book outside, take a moment now to run your fingers over the page. If the sunlight is strong enough, the paper will feel warm to your touch. Not only is the sunlight that reaches us full of different colours, but it also exudes warmth. In fact, the warmth that you feel in sunlight is the result of a wavelength of radiation that is longer than that in the red end of the visible spectrum: infrared. We experience this wavelength of radiation not through our eyes but through the sensation of heat.

This is what is fundamentally important about the electromagnetic radiation coming from the Sun and the reason why we cannot live without it – it carries energy to us in its electro-magnetic waves. Energy can come in many forms but what unites them all is the capability they have to be used to do some-thing – such as move an object or accelerate particles. The ability of charged particles to move as they feel a force in an electro-magnetic field means that the field contains energy. Infrared radiation is merely the form that is most obviously detectable by us because it heats our bodies by causing molecules in our skin to vibrate, but whether visible or invisible there is energy in electro-magnetic waves. This is what gives them the potential to heat materials, move charged particles, drive photosynthesis and much, much more. In fact, plants have adapted to absorb light specifi-cally at the red and blue wavelengths to power the synthesizing of nutrients; the remaining green wavelengths are simply reflected, which is why plants are pretty much always that colour.

Return of the particle

By the end of the nineteenth century, it looked like humans' understanding of light was finally in order. The observation of how light behaves, the investigation of electricity and magnet-ism, and the resulting theory of electromagnetism provided us with an explanation of light as a wave. There was just one small outstanding problem about how light actually transfers all that energy it seems to be carrying around. Explaining how the Sun's infrared radiation is heating up the page in your book was not quite as simple as scientists had hoped.

If instead of hitting your book and warming it up, the light beaming from the Sun hits a special sheet of metal instead, it can be made to bump some electrons loose within that metal and an electric current starts to flow. This process is at the heart of all modern solar panels. But the electric current does not behave in a logical fashion. If you produce a low-energy current and then dramatically turn up the brightness of the light hitting it you suddenly get . . . no change. More electrons are produced, but the energy of each one stays put. Yet a dimmer light, of a larger frequency, will produce a higher energy current. This problem was solved with some clever mathematics.

A German scientist, Max Planck, had shown at the turn of the nineteenth century that radiation emitted by a hot and glowing object could only be described by assuming that it is made of individual packets of energy, not a continuous stream of waves. He saw that if you did the calculations for light as a wave, the results did not match what actually happened in the real world. The only way to get the theory to match observations was to use a mathematical model based on light being made up of tiny discrete particles. We are back in Newton territory.

At the time, Planck dismissed the interpretation of radiation coming in packets of energy as nothing more than a mathematical manoeuvre to make the equations fit the observations – a convenient mathematical approximation and not a description of reality. But someone else did take the particle theory of light seriously: Albert Einstein.

In 1905 Einstein demonstrated that when radiation is absorbed, it acts as if made of individual particles, or photons. This was an important contribution to the Nobel Prize that was awarded to Einstein in 1921. The energy carried by each individual photon was related to its frequency – the higher the frequency the higher the energy, and two photons of the same frequency will always carry the same amount of energy. This explains the photo-electric effect: more light would not mean more energetic electrons as each photon is a discrete particle with an energy set by its frequency, and they all only knock out one electron each. Photons of higher frequency, which carry more energy, are needed for more energetic electrons.

It seems slightly unfair that, just as we began to understand light as a propagating wave, we should discover that light behaves as if it were being transmitted by particles. Thank goodness Hooke wasn't alive to see this. Newton, however, would have loved it. The wave and particle descriptions of light may seem counterintuitive but light is what it is, it behaves in its own way. It is up to us to find a description that best fits its behaviour and, for now, we need both. Part of the problem comes from our conceptual approach as we try to describe light in familiar and simple terms. Throughout this book both wave and particle descriptions will be used and I hope you'll feel light's seemingly fickle character starting to have purpose.

For now, one way to reconcile this apparent contradiction is to think of a beam of light as actually being made of countless individual flashes of light, which we call 'photons' – as if a torch were being switched on and off very quickly. The wave is then an average over the countless flashes of light.

As you sit in the sunlight, this energy is continually falling on you thanks to the vast number of photons of all different wavelengths that the Sun continuously emits. Every square metre of the Earth is being bombarded by 100 billion billion photons every second, which together deliver 1000 joules of energy – about the same energy as is used every second by a microwave oven. And over the sunlit side of the 510 million million square metres of the Earth's surface this really adds up. Some of this energy is reflected back into space but some of it goes into heating our oceans and atmosphere and powering modern society. Our knowledge of all this came from understanding what light actually is. Maxwell's work can perhaps be considered the greatest scientific achievement of the nineteenth century.

So the question now is: how does the Sun create sunlight? And where does all this energy come from?

2. Star Power

The Sun arises in the East,
Cloth'd in robes of blood and gold;
Swords and spears and wrath increase
All around his bosom roll'd
Crown'd with warlike fires and raging desires.

William Blake, 'Day'

Looking back into antiquity, we see two themes linked with the Sun: fire and power. In his poem 'Day', the English poet William Blake takes them up, talking about the Sun in terms of a warlike and menacing fire and anthropomorphizing our star as a warrior king with a flaming crown and powerful feelings. Well before Blake, the Aztecs, who lived in central Mexico 500 years ago, had Xiuhtecuhtli, their deity of both day and light and of fire. The Chinese had their Sun god, Yang, also made of fire.

Perhaps fire, flame and burning feature somewhere in your visceral feelings about the Sun too. From our everyday experience of the Sun's light and warmth, it's hard not to think of fire when we try to describe and understand the Sun. It might surprise you to know that this way of thinking isn't reserved just for the arts: scientists have thought along these lines too. The defining aspect of our Sun is that it shines – it makes its own light. And it's hard not to equate this with the most familiar example of light and heat we have here on Earth: a roaring fire.

The idea that fire is the reason why the Sun shines was considered as recently as the early nineteenth century. Against the backdrop of the Industrial Revolution, it was pondered whether a large enough coal fire might be able to provide the vast amount of light we receive from the Sun. It's an alarming thought: once all the coal was burnt, the Sun would go out; perhaps an early end-of-the-world scenario. But that was the same insight we could use to test the theory: the Sun has been burning for a long time and it hasn't gone out yet. How much coal would it have required so far?

To even begin to answer that question, scientists needed a coherent theory of energy. There is energy stored in coal, and by burning it that energy is freed and somehow converted into light and heat. But to even discuss energy as a quantity that can be measured, and as something that can be used, requires some serious advances in thinking. Thankfully the energy-hungry Industrial Revolution provided the motivation to achieve an understanding of energy. Unfortunately, the first step towards this understanding sounds hugely antiquated, and perhaps even racist, to our modern ears.

Julius Robert Mayer was a German doctor and amateur scientist. In 1840 he was travelling to Java, working as a ship's surgeon, when he had a strange epiphany. As he went about a nineteenth-century physician's duties such as letting blood, he noticed that the blood in the veins of the sailors who lived in warmer countries was more vividly red than that of his fellow Germans. He also relied on fire for his explanation of energy. He hypothesized that when food was 'burnt' in the body, it produced dark ash that went into the blood. If the body burnt more food, it would produce more ash and the blood would appear a darker red. Mayer speculated that sailors from warmer countries had vividly red blood because their bodies didn't need to use as much food to keep warm, and so didn't produce as much ash in the blood as sailors from colder countries.

We would find Mayer's physiological reasoning faulty today, but back then the analogy of fire equating to energy and warmth prevailed. The stroke of genius came when Mayer started thinking about other ways for the body to keep warm: for example, by rubbing your hands together. This gives the same result as a situation Mayer would have been familiar with on ships: when a rope runs through your hands it causes rope-burn — friction causes things to get hot. And, most obviously, physical labour, like swabbing decks and whatnot, would warm someone up too. By observing this line of cause and effect — food provides warmth for the body and so does doing work — Mayer arrived at a radical conclusion: mechanical work and heat were both being derived from a single quantity — energy — and this quantity could be transferred from one place to another but was always conserved. The overall amount of energy didn't change.

In England, meanwhile, James Prescott Joule, the son of a wealthy brewer, was working in Manchester, the city at the very heart of the Industrial Revolution. And central to the Industrial Revolution were investigations and experiments looking at heat and energy. The units we now use to measure energy bear his name, so it lives on all around us. I have a bag of corn chips on my desk as I write, and on the back it says that each 100 grams contains 2035 kilojoules. This is the amount of energy that can be extracted from the chips by my body when I digest them. I know that each 100 grams of chips contains 2,035,000 joules of energy. A frightening thought!

Joule's experiments led him to the same conclusion as Mayer, although they had very different approaches. In one experiment, Joule constructed a mechanism where a gradually falling weight pulled on a rope that moved a paddle and stirred some water. As the water was stirred its temperature went up: the mechanical energy of the paddle (derived from the potential energy of the weight) was turned into heat, showing that the mechanical energy of the paddle was equivalent to the energy of the heat. Again, energy had been transferred from one place to another.

The work of these two men was key in establishing a law that has become known as the First Law of Thermodynamics, which has at its heart the conservation of energy: energy can be neither created nor destroyed. Instead, it can only change from one form into another. This development in scientific under-standing was so important that a bitter dispute took place between the leading scientists of that time about who could take the credit for its discovery. They argued publicly; it was a matter of national pride with Joule and Britain being set against Mayer and Germany. In what could be seen as a conciliatory gesture the Royal Society awarded the most prestigious medal to Joule in 1870 for his 'experimental researches on the dynam-ical theory of heat' and to Mayer in 1871 for his 'researches on the mechanics of heat'.

Nineteenth-century experimentation also showed that to raise the temperature of 1 gram of water by 1 degree Celsius requires just over 4 joules of energy. Knowing this meant we could start to measure how much energy the Sun is producing. If you leave a known amount of water out in direct sunlight when the Sun is directly overhead and measure how long it takes to raise the temperature you can use this formula to calcu-late the rate at which the Sun's radiation transfers energy to heat the water. By the late 1830s, this had been done independently by the French physicist Claude Pouillet and the English astron-omer John Herschel.

John Herschel became interested in the Sun's energy for rather personal reasons. In a tradition maintained by English tourists overseas to this very day, he was severely sunburnt during a trip across the Alps in 1825. John Herschel was also the son of the famous astronomer William Herschel (who discovered Uranus*) and he was a successful astronomer in his own right. He took up astronomy rather late in life, continuing his father's work when he became too frail to do the work himself. After almost twenty years of working in astronomy, John then moved his family to South Africa to survey the skies of the southern hemisphere, where he catalogued stars and nebulae. He even made observations of Halley's comet when it became visible in 1835 and he realized that the heat of the Sun was causing cometary material to vaporize.

But Herschel wasn't the only person interested in the amount of energy delivered in sunlight. Claude Pouillet developed a new instrument that could measure the energy delivered in sunlight. He called this new instrument a 'pyrheliometer'. Herschel and Pouillet's experiments provided a way to measure the amount of energy falling on the very small surface area of their vessels of water. This value could be scaled up to find the energy falling on every square metre of the Earth. But it isn't the amount of energy that is falling on ground that is important if we want to calculate how much energy the Sun is emitting. Sunlight only reaches the surface of the Earth after having passed through the atmosphere, and that reduces its intensity. The amount reaching the upper limit of the atmosphere is more than that measured on the ground – their calculations gave a figure of 1260 joules per second per square metre.

Nowadays we have the ability to put instruments on satellites and make measurements above our atmosphere directly, and it

---

* In fact, William Herschel proposed that the Sun's output could be measured and made an attempt himself to do this qualitatively by looking at variations in solar output by studying how the price of wheat varied. He reasoned that when the Sun's output was low, crops would be less abundant and the price of wheat would go up!

turns out that Herschel and Pouillet were not too far off the actual figure of 1361 joules per second per square metre. In the units used by my corn chips, that's around 1.4 kilojoules. This is equivalent to about 0.1 grams of corn chips every second per square metre, which certainly does not sound like much. But this is just the rate of energy that falls on a relatively small area. To work out how much energy the Sun is emitting in all directions every second requires us to think big.

Imagine a huge sphere that it is centred on the Sun and extends out to the distance of the Earth, 150 million kilometres from its centre. Now visualize light flooding out from the Sun in all directions onto every part of the sphere. We know from the measurements at the Earth that every square metre receives 1361 joules per second, so to calculate the total amount of energy falling on the sphere we must multiply 1361 joules by the sphere's total surface area. This gives us an incredible value of 4×10^{26} joules every second (400,000,000,000,000,000,000,000,000 joules). The Sun produces the energy of around 20 million billion tonnes of corn chips a second. Because it is blasting out in all directions into space, this is over two billion times the amount we receive at the Earth. It is the amount of energy that the Sun radiates every second and this is the amount that whatever process is powering the Sun must be able to provide. So now we are in a position to answer the question: could the Sun be powered by burning coal?

Coal is comparable to corn chips: 100 grams of my chips can provide approximately 2000 kilojoules, 100 grams of coal can supply 1000 kilojoules (but tastes substantially worse). Burning 1 kilogram of coal releases 10,000 kilojoules of energy, but, still, this is a minuscule amount compared to the amount that the Sun is emitting. Even the most powerful coal-burning power station today would have to run constantly for thousands of billions of years to produce as much energy as the Sun does in one second. When you know the numbers, coal starts to look like a pretty measly energy source. Clearly a large amount would be needed to allow the Sun to shine so brightly. And, not only that, but the Sun would have to be massive enough to keep itself shining this brightly during its lifetime. Which brings us to an interesting detour: how massive would our theoretical coal-burning Sun need to be? And is the Sun bigger or smaller than that? If the Sun is smaller than that, we can rule that fuel out. But calculating the Sun's mass is not a short detour.

Critical mass

Calculating the mass of something as distant as the Sun is no easy feat. The first step in the right direction is thanks to our old friend Isaac Newton. Not just content with shining light through prisms and arguing with Hooke, he is also famous for discovering the law of gravity, apparently after having been inspired by an apple (150 kilojoules per 100 grams).

In 1687 he published his *Philosophiæ Naturalis Principia Mathematica*, which laid out a set of equations describing how objects move when they are and aren't being acted on by a force, and how an object that is stationary will stay at rest unless it is moved by a force. This includes celestial bodies, such as the planets, and how they move under the force of gravity. This describes yet another force acting at a distance, highly controversial at the time, which pulls two objects together with a strength that is dependent on their mass and separation from each other and nothing else.

As fundamental forces go, gravity is pretty weak. But the beauty is it works over vast distances. Newton had been inspired by the work of an earlier German mathematician and astronomer, called Johannes Kepler. In 1619 Kepler had published his finding that the time it took a planet to orbit the Sun is related to its distance from it. Newton expanded Kepler's observational work using his newly developed theory to provide a description of how planets move that involves the mass of the Sun and the mass of the planet. These theories take us most, but not all, of the way to working out the mass of the Sun. We need to put some numbers into the equations.

In fact, only one number is needed: the distance of a planet from the Sun – any planet will do. Despite having all the necessary equations in the seventeenth century, Newton didn't have this one numerical key that would unlock the problem of the mass of the Sun. It took until the eighteenth century for scientists to be able to calculate how far the Earth was from the Sun, but the delay wasn't in waiting for someone like Maxwell or Joule to come along and find a solution – we were waiting on the Sun itself, specifically that rare celestial occurrence: the transit, from the Earth's point of view, of Venus in front of the Sun.

We see a transit of Venus approximately every 115 years, when a pair of transits occur eight years apart. The first transit of Venus ever to be witnessed by human eyes didn't happen until 1639. Planetary motions were understood well enough by then to be able to predict that year's transit and two people in the know in Britain had skies clear enough to see it. By 1716 the British astronomer Edmond Halley (more famous for his work on cometary orbits) had worked out how to use a transit of Venus to calculate the distance to the Sun accurately. But the next transits of Venus were not until 1761 and 1769. Halley knew that he would not live to see these events and urged that the message be passed down so that astronomers alive then would make observations. Thankfully they did and calculated the distance between the Sun and the Earth to be between 92 million and 96.1 million miles.

Then, a small amount of number-crunching later, it was

calculated that the Sun has a whopping mass of 2×10^{30} kg, which is 1000 times more massive than all the planets put together and 333,000 times the mass of the Earth. Knowing that one kilogram of coal releases 10 million joules of energy, we can calculate that a pile of coal the size of the Sun would release 2×10^{37} joules. At the current rate the Sun is blasting energy into space, that stockpile would only last for 6300 years.

Despite the persistence even today of a small 'young Earth' cult, we now know that the Sun has been around for 4.6 billion years. There is no flammable fuel that could possibly be keeping the Sun going over this kind of time period. So if the energy source of the Sun is the very material of which it is made, and it is made of coal, it is simply not massive enough. Despite the Sun being synonymous with fire for all of human history, we can now say that the Sun is definitely not on fire. In the mid-1800s a more plausible theory was popular – that comets and asteroids falling into the Sun provided the energy source as the energy of motion of the incoming bodies was converted to heat. This idea fell by the wayside, though, when, despite over two centuries of telescopic observations, no comets or asteroids had been seen raining down on the Sun. A related alternative theory was that the Sun was shrinking and releasing energy from its own infall. The surface of the Sun would need to drop by just 35 metres per year to provide enough energy but this process could only sustain the Sun for around 20 million years. Far short of the 4.6 billion that we now know the Sun needs.

It took until the twentieth century for a theory to emerge that could cope with the amount of energy that the Sun needs to generate, kick-started by none other than Albert Einstein, who had been keeping busy since his work on the photoelectric effect.

Cannibalistic Sun: a little mass goes a long way

In 1905 Einstein published the famous equation $E = mc^2$. E stands for energy, m for mass, and c for the speed of light in a vacuum.* His incredible insight was that energy and mass were interchangeable. Einstein's equation shows that even a minuscule amount of mass can be equivalent to an awesome amount of energy, because the equation includes the speed of light, squared. So we have gone from releasing energy from a substance by burning it, whereby 100 grams of coal releases 1 million joules of energy, to Einstein's saying that, in fact, inherent in the mass of the coal is a much larger quantity of energy. A mass of 100 grams is equivalent to 10 million billion joules of energy in Einstein's formulation.

To provide the 4×10^{26} joules of energy that the nineteenth-century astronomers had calculated as being emitted in sunlight every second, the Sun would need to lose a few million tonnes of mass. But given the Sun has over a billion billion billion tonnes of mass to start with, it can certainly handle that. This line of reasoning is very appealing and gives another avenue to study now that our terrestrial fuel has hit a dead end. Is the Sun able to convert its own material into pure energy?

Perhaps. It's one thing to show there is theoretically enough mass in the Sun to provide this energy. It is much harder to show a plausible mechanism for actually getting that energy out. For the coal hypothesis this was relatively straightforward: you burn it and the energy is released during a chemical reaction. To release the full amount of energy that Einstein proposed was in the material would require its complete annihilation. The various inanimate objects around you right now may all contain billions of joules' worth of energy in their mass, but it (thankfully!) stays safely locked away.

* Although there is some evidence that other scientists were close to realizing this equation, e.g. Friedrich Hasenöhrl.

The first elements of practical support for Einstein's theory came at the start of the 1920s, when experiments to find the mass of atomic nuclei were being carried out. These experiments weren't being done to test Einstein's ideas, but to characterize the nuclei of different elements and understand what they were made from, but it was quickly realized that they provided some real-world support for what Einstein had proposed. The experiments showed that the mass of a helium nucleus was very slightly less than the mass of (what were thought at the time to be) its constituent particles. One helium atom somehow had less mass than the four constituent particles weighed separately. The mass of the whole nucleus was less than the sum of its parts and this mass difference could be explained and understood as energy liberated when a helium nucleus was formed.

Thanks to a British astronomer, Cecilia Payne-Gaposchkin, we know the Sun has plenty of hydrogen and helium. We'll be finding out about her work, and the challenge she faced in having this discovery accepted, in the next chapter. So in theory, then, perhaps a helium nucleus could be built from a hydrogen nucleus by a process of 'nuclear fusion' – joining together small nuclei of atoms to build bigger ones – and could release the energy everyone was looking for. This led to great excitement, but with one major stumbling block: protons are electrically charged particles that repel each other. Even at the high temperatures inside the Sun, where protons would be moving fast and colliding with each other at great speeds, there still wasn't enough energy to overcome the powerful electric repulsion of the two positively charged protons – at least, according to our old understanding of sub-atomic particles.

In 1928, though, the Russian theoretical physicist George Gamow came up with a description of sub-atomic particles that showed it was possible for charged particles to undergo a reaction at energies (that is, temperatures) lower than previously thought. Gamow's formula showed that on rare occasions charged particles would 'tunnel through' the repulsive barrier created by their having the same electric charge and be able to fuse together. This brought helium fusion back into the race as a legitimate candidate to power the Sun.

The final piece of the puzzle was provided by a German nuclear physicist, Hans Bethe, who started his career in Germany but ended up working in America, having moved first to Britain and then to Cornell University in 1934. His mother's family were Jewish, and in 1930s Germany that made him unemployable. When Hitler came to power, the Nazi government actually introduced a law that forbade anyone with a Jewish family from having a government-funded job. Bethe's research contract was terminated as a result, but this had the benefit of causing him to leave the country while he still could.

This was a significant move not just for Bethe's career, but for the course of the Second World War. His new home of America had no problem putting Bethe's genius to work and he ended up being recruited in 1943 to lead the theoretical division of the Manhattan Project at the Los Alamos Laboratory. Nuclear physics was the 'new' science that looked like it would be able to reveal the structure of matter. And with that would come a mastery of the material around us so that its energy could be released in the most catastrophic way. This led to a perceived race to produce the first nuclear bomb, and Los Alamos's single purpose was to do just that.

The list of people who worked on the Manhattan Project reads like a who's who of scientists whose names are well known today: J. Robert Oppenheimer, Richard Feynman, Niels Bohr, Leo Szilard and Enrico Fermi. It was a joint letter by Szilard and Einstein to the President of the United States at that time, Franklin

D. Roosevelt, which led to the creation of the Manhattan Project in the first place. They were concerned that developments in nuclear physics in Germany and German access to the necessary materials might mean that an atomic bomb was being built. Bohr had been awarded the Nobel Prize for Physics in 1927; Fermi had received the same award in 1938 – the Manhattan Project brought together the very best.

Despite the Earth-changing use for nuclear physics, Bethe's interest in fission and fusion started with his theoretical studies of atomic nuclei. In the late 1930s a conference that brought together nuclear physicists and astrophysicists turned his attention to the Sun. He set out to calculate whether it was theoretically possible for hydrogen to be fusing into helium at the heart of the Sun at a rate that would explain the Sun's radiated energy. The work of a British astrophysicist, Arthur Eddington, on stars as spherical balls of gas meant a temperature of around 40 million Kelvin in the centre of the Sun. This temperature came from Eddington's erroneous assumption that the Sun is mostly made of iron. Using hydrogen as the dominant element gives a central temperature of 12 million Kelvin. This was the temperature that Bethe used. Today we have upgraded this temperature to 15.6 million Kelvin. Rounding down gives this book its title. This is perhaps the most important temperature in the Universe.

Bethe knew that at its centre the material was incredibly dense too: 150,000 kilograms per cubic metre, which is 150 times the density of water and about ten times the density of lead. Eddington is central to our story here as, in fact, it was Eddington who immediately realized fusing hydrogen to helium might be the energy source of the Sun when it had been discovered that the mass of a helium atom is less than the sum of its parts.

At the huge temperature in the centre of the Sun, protons and electrons are so energetic that they break free of the electric charge that holds them together in the hydrogen atoms. Released from their bonds, the protons would be moving at over 500 kilometres per second. In turn, their motion would create a gas pressure at the centre of the Sun that is 2.5 billion times that of the air pressure around you. At such a high density and fast speeds, collisions between protons are very frequent. Armed with the knowledge of the extreme conditions inside the Sun, Bethe set about finding out how often Gamow's tunnelling could be taking place to produce a nuclear reaction (how many collisions are successful) and whether collectively these reactions could produce enough energy to power the Sun.

Bethe's work was show-stopping. He was able to demonstrate that the source of the Sun's energy is dominated by a series of events at the Sun's core that do indeed convert hydrogen to helium. But not in a straightforward manner. Known as 'the proton–proton chain', the path from hydrogen to helium is long and winding.

The first step takes place when two protons tunnel through the barrier they feel between them and a reaction takes place: one of the protons transforms into a neutron, which has no electric charge, and an antimatter particle called a 'positron' (essentially a positively charged electron) is formed along with a particle known as a 'neutrino'. The neutron does not hang around though, and combines with the other proton to form a new particle called a 'deuteron' (which now contains one proton and one neutron). The positron is also short-lived and rapidly encounters an electron – its matter equivalent – and the two annihilate each other, releasing energy in the form of a gamma ray in the process.

What's needed next is for the deuteron to fuse with another proton to form an almost-helium nucleus, which has two protons but only one neutron (whereas a fully formed helium

STEP 1:
proton fusion + deuteron formation

IN: 2 protons

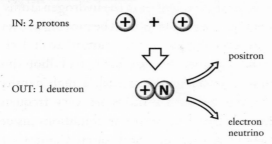

OUT: 1 deuteron

positron

electron
neutrino

STEP 2:
deuteron + proton fusion

IN: proton + deuteron

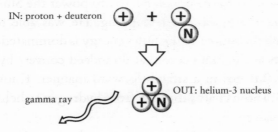

gamma ray

OUT: helium-3 nucleus

STEP 3:
helium-3 fusion

IN: 2 helium-3
nuclei

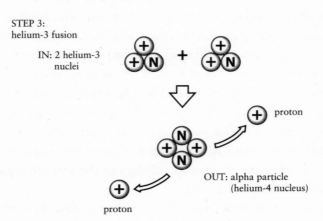

proton

OUT: alpha particle
(helium-4 nucleus)

proton

2.1 Flowchart illustrating the three basic steps in the proton–proton chain. Steps 1 and 2 need to occur twice for step 3 to be possible.

nucleus has two neutrons). Another gamma ray is formed too. Only then can the third and final step take place, in which two* of these helium nuclei come together to form a complete helium nucleus with two protons and two neutrons, leaving the two extra protons to go their separate ways. This means that, overall, four hydrogen nuclei have gone in and one helium nucleus comes out. Crucially for our world (and this book), the mass of the helium nucleus is 0.7 per cent less than the mass of the four protons that went in, and along the way this mass has been transformed into energy. And part of that energy, released in step two of the chain and in the electron positron annihilation, takes the form of a photon.

In recognition of his calculation which showed that this journey from hydrogen to helium was powering the Sun, in 1967 Bethe received a phone call. It was 6.15 a.m. but it would have been a very welcome call, despite waking him at such an inconvenient hour. As he answered, a Swedish voice at the other end of the line announced that he had been awarded the Nobel Prize for Physics – for work which was so significant that this was the first time a Nobel Prize had been awarded for a discovery in astrophysics, despite the fact that Bethe was not an astrophysicist and had done the work twenty-eight years earlier.

Is the Sun a nuclear bomb?

No. It's not.

The processes that occur inside the Sun may sound very similar to what happens in a nuclear (hydrogen) bomb, and much of the science was developed by the same people, but there are some fundamental differences.

* Necessarily, then, steps one and two must be completed twice for there to be sufficient particles for step three to take place.

While the aim of a thermonuclear bomb is to release a vast amount of energy in a very short amount of time, the Sun releases its energy relatively slowly and constantly. In fact, each time a proton–proton chain reaction is completed, the amount of energy liberated is tiny: a mere 4000 billionths of a joule. It is the sheer number of fusion reactions that take place in the core every second to power the Sun that leads to the vast energy output. If you could scoop up a coffee cup's worth of material at the centre of the Sun a billion reactions would be happening in the material every second. Combined together they would release just four thousandths of a joule. It would be like eating those 100 grams of corn chips slowly over sixteen years. Gram for gram the sailors studied by Mayer are able to create more energy from their food than the Sun does from its nuclear reactions.

On top of this, despite the enormous number of nuclear reactions that do take place in the core of the Sun, they are still only a minute fraction of the total number of particle collisions that occur. (Really, it's incredible that the Sun generates energy at all.) What counts is, again, the vast number of collisions that take place. What allows the Sun to produce the energy it does is its size: the Sun's core, where the conditions make nuclear fusion possible, occupies the inner 25 per cent of the Sun – a sphere 350,000 kilometres across, around thirty times the Earth's diameter. This is so large that, despite the low likelihood of a collision resulting in a fusion event, the huge number of proton collisions really add up: every second, 600 billion kilograms of hydrogen in the core are turned into 596 billion kilograms of helium, with the missing 4 billion kilograms being turned into other forms of energy. Luckily for us, mass is not something the Sun is short of and it has enough hydrogen in the core to keep it shining for some billions of years yet. In fact, the Sun is a middle-aged star – it has been fusing hydrogen to helium in its core for around 4.6 billion years but has enough supplies to live out another 4.6 billion years, meaning that it is only halfway through its life.

3. Suns and Daughters

Given the enormous number of nuclear reactions that take place in the core of the Sun, why doesn't it explode now? There is a vital difference between the Sun and a nuclear bomb: there are 522,000 kilometres of gas lying between the core and the surface of the Sun and this gas tames the thermonuclear bomb within it, turning it into a clean energy source with a power output and longevity that we currently can only dream of re-creating on Earth.

All that extra material around the Sun's core does throw up two new mysteries though. Firstly: what is it? Is it the same material as that fusing in the Sun's core? To answer this question we have to look at how the Sun was formed.

That the Sun does not explode like a thermonuclear bomb is because of a delicate balancing act, one that keeps the Sun exploding enough to stay hot, but not so much that it blows apart. We now know this equilibrium was established when the Sun was young and that fusion proceeded at a steady rate, and has remained safely stable ever since. For centuries scientists have been trying to discover how the Sun formed, with no idea they were also working towards a theory of how it keeps its thermonuclear reactor in check.

What we see now when we look up at the Solar System around us is just one snapshot in its life cycle. The Solar System has been here for almost 5 billion years and has about that long left to go. As a family, the Sun and the planets have been through a lot together already, and they have plenty of ups and downs still ahead of them.

If the Solar System were an actual family, the Sun would undoubtedly be the boss, but I'm not just saying this because I have made a career out of studying the Sun. It is the dominant member of our cosmic gang. It is a single parent trying to raise its offspring planets in a modern galaxy. It is the head of the family.

But even though the Sun rules the roost and makes up over 99 per cent of all matter in the Solar System, that 99 per cent may be the most difficult to study. Historically it has been easier to track the movement of the planets in the sky and to try to understand them. But studying all family members, even looking at planetary motions, can help us to understand the Sun. The first realistic family portrait of the Solar System was formed in 1755, by a German philosopher, Immanuel Kant.

As well as being a hugely influential philosopher, largely responsible for the Western philosophy of the past three centuries, Kant also dabbled in science. This was at a time before academics had the clear-cut roles we do today. In a modern university you would not expect to see a solar physicist striding into the philosophy department! Actually, to be more accurate, the era that Kant worked in was a time when modern science was gradually breaking away from its philosophical roots – though the connection lingers today: I have a Ph.D., which is the abbreviation for Doctor of Philosophy, in solar physics.

In Kant's time, the Solar System was thought of as a small family. Only five planets other than the Earth were known – Mercury, Venus, Mars, Jupiter and Saturn, as they can be seen without the aid of a telescope. The complete family portrait actually includes Uranus, Neptune, Pluto and friends, the Kuiper belt objects, asteroids and comets. But that picture was still to be painted. And we can't be smug even now as we are still working on the portrait: new members of the family are still being discovered. The information that Kant needed was all there, though.

Kant used not only the movements of those five planets but also the way they spin to create his picture. The Earth rotates once every twenty-four hours, giving us our day, but it is not unique in that: all of its siblings spin as well. What struck Kant were the striking similarities in how all of the planets spun and moved: they all orbited the Sun in the same direction and they all spun in that same direction. If you were to look 'down' on the Solar System (from a northern-hemisphere-centric point of view) all of the planets went around the Sun anti-clockwise and they all spun anti-clockwise. Not only that: they didn't orbit the Sun in a haphazard fashion, but all moved on one plane, as if they were all rolling around on the same flat surface.

It should be said, though, that today we know things aren't quite as simple as this. Two planets spin in the opposite direction to the rule Kant stated: it was not until the 1960s that radar measurements discovered that Venus, which is shrouded by thick clouds, spins the opposite way; as too does Uranus, a planet not visible to the naked eye and not known about by Kant. But these are exceptions that still fit within Kant's theories with some modern tweaks.

Because all the planets share the same family traits, Kant knew they must have all formed together – the planets were not created individually and then later adopted by the Sun but must have all originated via the same process. Kant wound the clock backwards and concluded that the Solar System must have started as the one nebulous cloud of gas that then gave birth to the planets and Sun we see today. The sticking point with this is that his theory predicts that the Sun and the Earth are made of the same stuff. But it is hard to think of two more dissimilar objects: the water-soaked Earth and a raging ball of super-heated gas. To test Kant's theory requires us to know more about what the Sun is composed of.

What is the Sun made of?

The answer to the question, 'What is the Sun made of?', was finally cracked at Harvard College Observatory (in Cambridge, Massachusetts) in the 1800s and 1900s. Cambridge is still a focus for research, and the Observatory archives, part of the Harvard–Smithsonian Center for Astrophysics and which I visited while on a research trip, is an incredible place where almost 200 years of astronomical history are encapsulated – and it's also a very significant place for women in astronomy.

In the nineteenth century the staff at the Observatory were given the enormous task of categorizing the entire night sky. This wasn't a simple matter of photographing every star – the light from each star was passed through a glass prism and the resulting spectrum recorded. It would take a huge amount of work. Worldwide, around 6000 stars are visible to the naked eye if the skies are dark enough. Using just a small telescope raises this number into the hundreds of thousands. To get through all of these stars, Harvard College Observatory hired a team of women.

They became known as *computers*, well before a 'computer' changed from being a person to a machine. These women computers had the task of effectively processing the astronomical data and categorizing the stars according to any similarities they showed in their individual spectra. Sadly, they were not hired to do any actual research themselves; their job was to do the tedious work so that the real scientists didn't have to. But these women were not going to be held back by a mere job title. They did some amazing work in their own right.

During my visit I walked past cupboard after cupboard that together house over half a million photographic glass plates showing the light captured from hundreds of thousands of stars. These plates are the actual data that the computers would have worked with. I wanted to see the glass plates studied by one computer in particular, Annie Jump Cannon. Cannon was a phenomenally efficient and talented computer and categorized over 250,000 stars during her career, sometimes at a rate of three per minute. She would sit near a window and use a mirror to bounce the incoming light through the glass plate, decipher what she saw and enter the correct classification into a ledger before moving on to the next. (See plate 1.)

Cannon's classification of stars led to a system in which stars were organized by their temperature. This classification system is still used today: OBAFGKM. Stars range from 'O'-type stars with surface temperatures from over 30,000 degrees Celsius (Kelvin, to be more specific) to 'M'-type stars with surface temperatures of around 3000 Kelvin. Because of the erratic ordering of the letters, I was taught a mnemonic at university to help remember the scheme that Cannon had devised – 'Oh! Be A Fine Girl – Kiss Me!' Or, as I prefer to remember it, 'Oh! Be A Fine Guy – Kiss Me!' I was able to see the desk at which Cannon sat and the plates that she studied. In recognition of her outstanding talent, the American Astronomical Society now makes an annual award in her name to an early-career female astronomer who shows exceptional promise.

Cannon was able to classify stars based on spectra because scientists had learnt how to decipher all sorts of information encoded into these little rainbows of colour. What Newton had seen as a range of colours was far from being that simple. By looking at a spectrum it is possible to tell not only how hot the source must have been, but also even what it was made of. The Sun and other stars could easily have remained for ever out of humankind's reach, but it turns out that their light has brought all the information we need right to our doorstep.

Gaps in the rainbow

If you take the same rainbow that Newton saw and spread it out over a much greater distance you will start to see finer detail. It is not a continuous range of colours: there are gaps. If you look closely, there are dark lines within the Sun's spectrum – thin slivers of colour are completely missing.

These gaps were first seen by an English doctor, William Hyde Wollaston, in 1802. He was studying how light bends when it shines through various substances (the refractive index) and he happened to cast a very wide spectrum, and spotted the dark lines. He only saw a few and decided they must be the gaps between colours. In 1814 these lines were independently rediscovered by a German optician, Joseph von Fraunhofer, but with his superior optical equipment he found hundreds of them. Today these dark lines are known as the 'Fraunhofer lines' and thousands have been identified. Fantastically, these Fraunhofer lines are caused by the Sun and are not some new property of light. Understanding how they are formed goes as follows.

If a solid or a gas is heated enough it will start to glow. This is a familiar concept and gives us the phrases 'red-hot poker' and 'white heat'. It's the same principle behind incandescent light bulbs too: a tungsten filament is heated to over 3000 degrees Celsius and it starts to glow. If you looked at the spectrum from a hot-glowing object, you would not see any missing lines though; the spectrum would be completely continuous.

Another way to produce light is to energize a thin gas. This is the effect that is utilized in neon lights and some energy-saving light bulbs. In this case, a gas is given energy by passing an electric current through it, not by heating it. But if you looked at the spectrum of a neon light you would get a shock. It is the complete opposite of the nice continuous spectrum from an incandescent blub. It would not even be a 'spectrum' in some senses of the word; all you would see would be a few discrete bright lines, each of a single colour, with swathes of darkness in between. (See plate 2.)

The light we see from the Sun looks as if a continuous incandescent spectrum had had these disjointed gas spectral lines subtracted from it. And it turns out this is exactly what has happened.

In 1859, a German physicist, Gustav Kirchhoff, made a discovery that brought this unclear variety of spectra into focus. Kirchhoff carried out a series of experiments using very pure gases that were made only of single elements, and as he analysed the light of each one he realized that they all have their own unique disjointed rainbows, what we now call an 'emission line spectrum'. The coloured lines produced by lithium are seen at different wavelengths to those of sodium, which are different to those of potassium and so on. It was no longer necessary to touch a substance, cut it up or do an experiment with it to find out what it is. As long as the substance you are curious about emits light, you can study its line spectrum and know exactly what it is. And the Sun is not short of light.

Kirchhoff's next discovery explained why these emission line spectra have been subtracted from the Sun's continuous spectrum. He showed that if a thin (and relatively cool) gas was placed in front of a light source that produced a continuous spectrum, the emission line spectrum reversed itself and the previously bright lines became dark. The gas was now absorbing light at those wavelengths, not emitting them. The continuous spectrum (formed by the background incandescent object) with dark lines in it (formed by the intervening gas) is known as an 'absorption spectrum'.

Kirchhoff realized that if the wavelengths of the dark lines in the Sun's spectrum were measured, they could be compared with the emission/absorption line wavelengths that he had found the elements in his lab produced, so that the chemical make-up of the solar atmosphere could be found. The absorption lines could be used as a cosmic barcode – a remarkable leap forward and one that totally changed the way in which the Sun could be investigated. Even without visiting the Sun, there was now a way to find out what it is made of.

It was these absorption lines that Annie Jump Cannon and the computers in Harvard were looking at. Not only could the spectral absorption lines from distant stars be matched with terrestrial spectral lines to tell us what elements must be present in those distant stars, but the particular lines that are present reveal what temperature those elements were at when they stole the photons from the spectrum. This was the vital work done by those women: they were exploring and sampling stars across our Galaxy.

This technique conclusively showed that the Sun contained the same elements as the Earth, and acted as a kind of maternity test. But there was a problem – something which showed that the Earth and the Sun must have had very different upbringings. And this was discovered by another amazing female, who was also at the Harvard College Observatory. This woman did achieve the status of being able to do her own research.

The discovery was made by a British astronomer, Cecilia Payne (her maiden name – she was later Payne-Gaposchkin), who began her scientific studies in 1919 at the University of Cambridge, where she had enrolled to read Natural Sciences, with a focus on botany. But then she attended a lecture by the director of the Cambridge Observatory, Arthur Eddington. He had just travelled to an island off the west coast of Africa to see a total solar eclipse and his lecture was on how his observations of stars close to the Sun in the sky had provided the first experimental proof of Einstein's general theory of relativity. Einstein had only published his theory in 1915 and it was

Eddington's observations that thrust Einstein into the international limelight. The early twentieth century was an exciting time for physics, and after the lecture Payne abandoned botany to focus on astronomy.

Unfortunately, Payne was studying at Cambridge several decades before women were awarded full university degrees, and there were certainly no opportunities for a research career in astronomy once she had completed her course. It was to fulfil her ambitions that she had to move across the Atlantic to Harvard College Observatory in 1923, to study for a Ph.D. in astronomy (literally moving from Cambridge to Cambridge). Going there gave Payne access to the observatory's vast collection of stellar observations and placed her amongst the greatest astronomers of the era.

Payne's aim was to use the Harvard data to study stellar chemical composition. As she intricately examined the spectral lines in the light from the Sun and other stars she applied the latest theoretical ideas about how the light they emitted was affected by their surface temperature. In 1925 Payne discovered that even though the Sun has the same elements as the Earth, its composition isn't at all similar to the Earth's: the ratios of the elements are completely different.

She found that the most abundant element in the Sun is hydrogen – the lightest and least complex of all elements in the periodic table. The next most abundant was helium. As for those elements commonly found on Earth – such as iron, oxygen and silicon – they accounted for less than 2 per cent of the Sun's composition. This was a highly controversial result and put Payne up against the establishment, who had accepted the idea that the Sun should have the same proportion of elements as the Earth. Even Eddington thought that iron was the most abundant element in the Sun.

Payne was forced to write off her findings as nothing more

than a discrepancy in the data. The biased opinions of an older generation and their strong expectations of what should have been observed meant this remarkable discovery was totally overlooked. Payne's gender and the lack of influence that women had in the astronomical patriarchy wouldn't have helped either. In the years following Payne's discovery, the work of established male astronomers confirmed her result and for a long time they took the credit. Payne received little acknowledgement for her discovery of the major constituent of the Sun, the stars and indeed the whole Universe.

However, Payne's role at the Harvard College Observatory did signal the start of a gradual change in both the role and the position of women at Harvard and in astronomy more generally. She carried out her own research and published her own ideas and pioneered the transition from women being employed only as computers to becoming researchers with ideas and leadership of their own. Even though she was silenced at first, she did eventually achieve recognition later in her career. She blazed a trail for generations of women who followed her, and her observations changed our view of the Universe, just like Eddington had done for her a few years earlier. It's always surprising to me that her name is rarely mentioned.

Pulling it all together

Right, let's put it all together and actually build our Solar System. We can deal with the differences in composition between the Sun and the Earth later. This is the theory first put forward by Kant:

In the beginning there was a vast cloud of gas and dust, far enough from any other stars not to be significantly affected by their gravity. This explains the similarities across the Solar

System; everything came from the same cloud of dust and gas. The cloud was simply drifting through our Galaxy, insubstantial and fragile, with only a few particles per cubic centimetre and stretching over a region millions of times bigger than the Solar System today. Then, for some reason, it started to collapse.

There is still some speculation about how and why this happened. My favourite contemporary theory posits that it could have been the consequence of a nearby exploding star, a supernova. The pressure pulse from the shock wave created by the explosion could have compressed the cloud sufficiently for its own gravitational force to draw it in on itself. No matter how the process began, though, the particles moved closer and closer together, imperceptibly at first but then ever more rapidly as their own collective and self-perpetuating gravity pulled them inwards.

As the cloud shrank it became denser, forming what is now known as the 'solar nebula' – literally, the cloud that became the Sun and the rest of the Solar System. But as the cloud collapsed down into a smaller and smaller region two strange things started to happen: it began to spin and it flattened out into a disc. Actually, it didn't begin to spin – the cloud had always had a slight rotation, though it was just too subtle to notice. But any overall rotation of the initial gas cloud, no matter how slight, became amplified as the nebula contracted. As the nebula shrank, its rotation became faster and faster, just like an ice skater drawing in their arms to make them spin more rapidly. This occurs because the angular momentum of the nebula – the product of its size, mass and rate of spin – must remain the same. This is the law of conservation of momentum. So if one component, like the size, goes down, another component, that is the spin rate, will go up.

This spin explains the flattened shape. As the material began to spin faster and faster, the material around the rapidly rotating

Sun flattened into a disc. The reason for this is that the spinning nebula was experiencing another force – other than the inward force of gravity. It's a force that we have all felt as children when playing on a roundabout or as adults driving a car round a sharp bend: we feel like we are going to be spun outwards – not up or down, but outwards from the centre. We know that spinning sends you in a very predictable direction and this happened to the nebula. As the nebula contracted, and started to spin more and more rapidly, the gas particles experienced a stronger and stronger outward force.

What happens next depends on where the gas particles are in the nebula. The outward force is strongest at right angles to the rotation axis – in other words, in the plane the nebula is spinning in. So gas particles in this plane have a harder time being drawn in as the outward force works against gravity. This creates a flattened disc. Imagine a pizza chef. He or she starts with a ball of dough, and once they start whirling the dough above their head it creates a thin but wide spinning dough-disc.

Within the swirling disc of the nebula, it's thought that particles started to collide and within these collisions some particles were able to stick together because of the electrostatic forces of the electrically charged particles in the nebula – minuscule fragments at first. And, as fragments started to form, some collisions would break them apart whilst others led to fragments joining together so that they gradually grew in size. Fragments then became clumps and, as they got massive enough, the attractive force of gravity became strong enough for clumps to start to glue together too. Gradually, over a few million years, these clumps eventually became large enough to form the planets. And as more and more planetary material coalesced, the regions in between them started to empty and, around 10 million years after the solar nebula began to collapse, the formation of the Solar System was well under way.

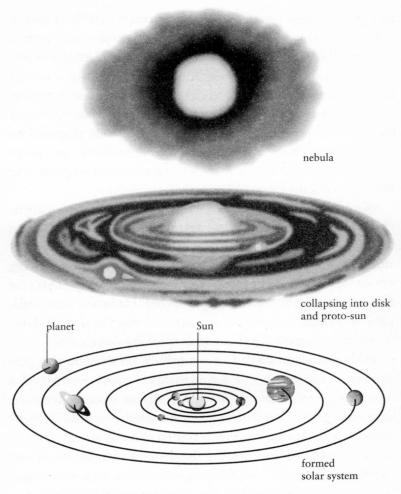

nebula

collapsing into disk
and proto-sun

planet   Sun

formed
solar system

3.1 Cartoon of the formation of the Solar System from the collapse of a
vast cloud of gas and dust known as a nebula. Not to scale!

The formation of the Solar System from the collapsing and
spinning nebula explains why we have a spinning Sun and spinning planets orbiting around it. They all came from a rotating
cloud of material and so they all inherited that same sense of

spin. But while some of the gas and dust was able to remain in orbit and form the planets, asteroids and other Solar System detritus, these are the outliers. A much larger amount of the dust became the Sun. Early on there would have been about an Earth's worth of matter in a clump, and then a Jupiter's worth. But more gas and dust kept raining down and the ball continued to grow. Eventually it became a sphere that was possibly 100 times wider than the Sun is today.

All this gravitational collapse had the effect of heating the matter up. In the same way in which an object dropped from a height to the earth accelerates as it falls, so the particles' increasing 'kinetic' (i.e. moving) energy came from their decreasing 'gravitational potential' energy (i.e. the amount of energy inherent in any mass that has not yet succumbed to gravity). Put plainly, the closer each particle got to the centre of the nebula, the faster and faster it was moving. And increasing the kinetic energy of all the particles increased the temperature of the gas.

Then this process of gravitational collapse carried on, as the massive proto-Sun started to crush itself down. The centre of this ball of matter was now starting to get extremely hot – so hot that it began to glow. This was the first stage of the Sun's career producing light. But it was only the warm-up round. About 50 million years after the cloud collapse began, the temperature in the centre of the Sun reached the point where, thanks to Bethe and Gamow's work, we know thermonuclear reactions would start. It was like billions of light switches had been flicked.

Importantly, the thermonuclear reactions stopped an endless gravitational collapse of the Sun. The collapse had been important because it compressed and heated the centre until it was hot enough – and critically dense enough – for fusion to take place. But the nuclear reactions were now giving the Sun a way to push back, and this held up the rest of the Sun and it ceased collapsing.

And this is the balance that keeps the nuclear reactor in the

Sun under control to this very day. If the fusion reactions were to suddenly increase (for some unknown reason), the core would heat up, the pressure would increase and the gas would expand. This expansion would have the effect of reducing the core temperature and density, which would have the effect of bringing down the reaction rate. But if the fusion does ever start to 'go out', this drop will allow the Sun to continue collapsing, raising the density and temperature and increasing the reaction rate again. It's this interplay and feedback between pressure, gravity and rate of fusion within the Sun that keeps it regulated, so that it doesn't explode like a hydrogen bomb releasing its energy in a short but catastrophic burst. It works exactly like the 'governor' in a car engine that allows it to idle at a nice constant rate.

But how do we explain the discrepancy between the chemical compositions of the Sun and the Earth? We still think that the solar nebula theory is a good description of what happened. The answer is that the Earth and all of the so-called 'rocky' planets that formed near the Sun – Mercury, Venus, the Earth and Mars – were too small and too hot because they are close to the Sun to hold on to the lightweight hydrogen and helium gas that would have been present in their atmospheres when they formed. At such temperatures, hydrogen and helium particles would have been able to escape the gravitational pull of these planets. Whereas the Sun has mostly maintained its original composition (except for the changes due to fusion, hidden from our sight in the core), the Earth has evolved. You can think of the Earth as having started the same as the Sun, but having since been distilled down to be almost only heavy elements.

For all the successes of the solar nebula model it still makes some predictions that do not match today's observations. According to the model, the Sun and the planets should all be spinning in the same direction. This nicely described the situation until the discovery that Uranus's axis of spin is at 90 degrees to that of the other planets and that Venus spins backwards. Similarly, one would expect, owing to the laws of momentum, that the most massive object in the Solar System, the Sun, which is at the centre of the spinning disc that is the Solar System, would be spinning at a greater rate than the much less massive planets further out. Here we find a serious problem: the Sun is spinning 400 times more slowly than we would expect. At some point during its 4600-million-year life, it must have transferred some of that momentum out into the Solar System. A mystery we'll revisit later.

Even with these few open problems, the solar nebula model remains the most likely scenario for the formation of our Sun. The most compelling evidence is that we've seen other nebulae in action. Elsewhere in our Galaxy we can watch younger stellar system families growing up. The most famous nearby nebula in which stars are forming is easily observable with binoculars and is found in the constellation of Orion. This constellation shows Orion the hunter, who carries a club and a shield and has a sword hanging from his belt. In the sword is a fuzzy region of nebular gases that have been lit up by newly formed stars.

The Orion nebula was observed as soon as telescopes were developed, with records dating back to 1610. Today it is known to be a vast cloud 40 light years across and around 1300 light years away. Many stars are forming in this vast nebula and the young ones have created strong winds that are blowing away the gas and dust from their surroundings, allowing us to peek into this stellar nursery and see what it is like to look back in time to the formation of our own Solar System.

Finally, the elements that make up the Earth and the Sun give us a glimpse of what came before our current family – back to previous generations. Modern measurements of the composition of the Sun show that, by mass, the solar nebula must have been 73 per cent hydrogen and 25 per cent helium, with all the other elements that we are more familiar with on Earth contributing only 2 per cent of the mass. Hydrogen and helium were created during the formation of the Universe itself, so their presence is understandable.

As for the other elements, the only way to explain their presence is if they were forged by nuclear fusion, something that can only happen in the centre of a star. Instead of reading this text as a book printed on paper, you might be reading it on an e-reader or a smartphone. In your hands now will be some very heavy elements that make up your electronic device, such as copper, aluminium and zinc. We know that these elements can only have been formed in stars far hotter than our own Sun is at the moment. And there are other elements such as tin or uranium that were formed when previous stars exploded as supernovae.

We now know that our Solar System formed from the remains of previous stars. They lived their lives, burnt out and scattered their ashes into the Universe. Billions of years later, the dust and gas from these ancient stars formed our nebula. And not only does our Sun have parents – we now know that it has grandparents. We live with a third-generation star, with a heritage reaching far back into the Universe. This third-generation star makes its own light by cannibalizing its own material. Not only that, but the photons released in the proton–proton chain reaction are of an extreme and rare variety here on Earth: gamma rays, with the very smallest wavelengths and one of the very highest energies. They are well beyond the visible part of the spectrum and are very harmful. But by the time those same photons reach the Earth, they have become relatively harmless sunlight. Something else is going on. What happens to all those photons as they pass through the rest of the Sun and eventually escape from its surface? Something serious must be happening because they change character completely.

4. The Secret Life of a Photon

We have learnt that the Sun stays hot and produces electromagnetic radiation by the process of nuclear fusion, which means we now know what light is and where sunlight is born – except that the high-energy, dangerous gamma ray photons produced by nuclear fusion are very different to the almost harmless, garden-variety sunlight we all take for granted on Earth. The mystery is how this transformation takes place.

They must change identity within the Sun itself, because the photons remain unaltered during their 8 minute and 20 second journey from the Sun to the Earth. Their transformation must occur as they travel from the centre of the Sun to the surface . . .

. . . a journey that takes a single photon around 170,000 years.

The light reaching you right now from the Sun began its journey when archaic *Homo sapiens* were still considering evolving into modern humans. The next time you are out in the sun, take a moment to appreciate that the light you are seeing is as old as our species.

As we know, even though light moves extremely fast in a vacuum (300,000 kilometres per second) it slows down when it passes through other substances. And whatever is happening inside, the Sun is slowing the photons down by an incredible amount. But it does not stop them completely: the photons still creep their way across the 522,000 kilometres of gas above the core before they reach the surface, at an average of 3 kilometres a year – this is less than half a metre an hour. So while the inside of the Sun is not completely opaque, it does make the photons' journey out into the Solar System a Herculean task.

The first model of the interior of the Sun was constructed in 1869, around the time that Kirchhoff was carrying out his experiments but well before Payne's discovery of the composition of the Sun and the realization that nuclear fusion was the power source. It was made at a time when even the physical nature of the Sun was still debated and the model was the work of Jonathan Homer Lane. Lane was an American scientist with a broad interest in science, engineering and experimentation. The best insight into the personality of Lane comes from the diary of a contemporary, Simon Newcomb.

Newcomb was a Canadian-American astronomer and had recently read an article in an English weekly publication called the *Reader*. This article suggested a new theory that the Sun could actually be a mass of incandescent gas, rather than a molten liquid as was the popular view at the time. He decided to tell a few of his friends from the local scientific club about it, including one strange person he described as 'an odd-looking and odd-mannered little man, rather intellectual in appearance . . . I did not even know his name, as there was nothing but his oddity to excite any interest in him.' Newcomb soon realized he was making that classic scientist's faux pas: accidentally describing a theory back to the person who came up with it in the first place.

Lane was of the opinion that the Sun was made of gas – a controversial view at a time when there were even some who considered it possible for the inside of the Sun to contain cool material that had condensed into a solid. If it were a gas, though, then Lane knew from the work of Lord Kelvin that the surface of the Sun was likely to be much cooler and less dense than its interior. Newcomb, having later become one of Lane's few friends, actually introduced him to Lord Kelvin and so began the wider acceptance of Lane's theory.

Lord Kelvin began his life as William Thomson, in Belfast, in 1824. But by the end of his life he had moved to Glasgow, made several landmark scientific discoveries, and been promoted to the title of 1st Baron Kelvin. Kelvin was the proponent of the theory that the energy source of the Sun was its own gravitational collapse. Occasionally, I have referred to 'Kelvin': units of the scientific scale of temperature named after him. These are exactly the same as degrees Celsius, except that they start 273.15 degrees lower. So a room at a toasty 30 degrees Celsius is at 303.15 Kelvin. Given that we round off temperatures in the Sun to the nearest thousand degrees, if not the nearest million, it actually does not matter if you use Celsius or Kelvin.

Lord Kelvin had been developing a theory about the Sun based on his thinking about the Earth's atmosphere – because physics is physics and the same laws apply across the Universe, so the Sun's gaseous mass might be understood by considering what we see here on Earth. In the Earth's atmosphere convection is common: warmer air rises while cooler air sinks. This is because the atmosphere is held in place by the gravitational pull of the Earth, so that cooler, more dense air has more mass to be pulled back down, displacing the warmer, less dense air. The same physics also applies in your kitchen and you will see an extreme version of this when you boil a saucepan of water: the low-density steam forming at the base of the saucepan rushes up to the surface.

Kelvin considered that there would be a temperature gradient across the air in the Earth's atmosphere. The convective motions do not disrupt this overall temperature structure; on the contrary, they maintain it. Any bubbles of air that are in motion will expand or contract until they reach the same density as their new surroundings. This alters the temperature of the gas bubble, and balances it with the surroundings.

By making an analogy with the Earth's atmosphere Kelvin had reasoned that the gaseous sphere of the Sun could also be in convective equilibrium, with layers of gas at different temperatures that get cooler the nearer they are to the Sun's surface, where the energy is leaking into space. From this starting point, Lane brought to bear data based on experimental work: the measured value of the brightness of the Sun and how hot materials emit light. He calculated values for the temperature and density of the gas inside the Sun, from the centre outwards, and arrived at an estimate for the surface gas temperature. This value was remarkably close, within a factor of five, of what we now know it to be.

These ideas developed by Lord Kelvin and Lane of the Sun being a ball of convecting gas, a massive roiling cosmic cooking pot, were in line with observations. The master of eighteenth-century astronomy, William Herschel, first noted a mottled appearance to the Sun in a paper he published in 1801. He wrote that 'Corrugations change their Shape and Situation; they increase, diminish, divide, and vanish quickly.'

The corrugations are now known as 'granules'; if you zoom in on the Sun it looks almost exactly like a pot of boiling water. You can see the mottled appearance as plumes of hot gas reach the surface – only, instead of bubbles a few millimetres or centimetres wide as you see in boiling water, these granules are about a million metres across. The spacecraft I work with send back some amazing videos of the surface of the Sun in motion. (See plates 3 and 4.) If you get a chance to see them yourself, perhaps through an observing event at your local astronomical society or through pictures online, they are a sight to behold.

Modern data show that there is more to the convection than just the granulation that Herschel saw. There are larger convection cells that engulf the smaller ones – like a series of Russian dolls. The largest convection cells, in features called 'supergranulation', stretch out over 30 million metres at the photosphere. Then, between the size scales of supergranulation and granulation, it has been proposed that an intermediate mesogranulation is circulating. But this has yet to be proved and the consensus at the moment seems to be that mesogranulation is nothing more than a misleading artefact in the data.

This may feel like an open-and-shut case, but sadly Lord Kelvin's and Lane's model is an incomplete explanation of the Sun – really it just scratches the surface. We now know that the 'convection layer' of the Sun only extends about 18 per cent of the way to the centre of the Sun. It turns out that there is only a very tenuous shell in which convection is occurring, and below that in the Sun things are very different indeed.

In the radiation zone

School students are taught about the three ways in which energy can be transmitted: convection, conduction and radiation. The top 18 per cent of the Sun is all about convection, and nowhere in the Sun are conditions really suited to conduction (conduction normally plays a role in solids). The way to transmit energy through the remaining 82 per cent of the Sun is by means of radiation. This radiation is comprised of the photons themselves that are trying to get free. The first realistic model of the radiation within the Sun was developed by Eddington, not long after he had confirmed Einstein's ideas about gravity and inspired Payne to switch from botany to physics.

'At first sight it would seem that the deep interior of the sun and stars is less accessible to scientific investigation than any other region of the universe'– this was the opening statement of Eddington's 1926 publication *The Internal Constitution of the Stars*, one of the most important astronomical texts ever written. In it Eddington abandoned the earlier ideas of energy transport within stars being achieved solely by convection and considered the role of radiation.

The reason why Eddington's work became so influential, and why it is central to our story here, is that by making radiation the primary means of energy transfer within the Sun he came up with a picture of it that explains, at last, the true state of the hydrogen gas within it and how and why it is able to affect the photons on their journey from the core of the Sun to the surface. What Eddington realized is that the hydrogen gas in the Sun would be in the form of 'plasma', as it is known.

Plasma is often referred to as the fourth state of matter, after the sequence: solid, liquid, gas. Take the water in your saucepan, for example. Below zero degrees Celsius all that water would be frozen solid; between zero and 100 degrees it is a liquid; above 100 degrees it boils as some starts to turn into a gas. But what happens if you trap the steam in your pan and continue to heat that gas? Would it go through another transition?

Without any high-pressure cheating, at normal atmospheric pressure steam will change again at around 12,000 degrees Celsius, turning into the fourth state of matter: a plasma. This is the point where the energy of the atoms is so great that the electric force holding them together is overcome. The water molecules break up into single hydrogen and oxygen atoms. And then the electrons in the atoms themselves would start to break away from their nuclei. This process, whereby a neutral atom is changed into an electrically charged ion, is called 'ionization', and when this happens to many particles at once, it produces plasma.

This is what happens inside the Sun. Strictly speaking, the Sun is not a ball of gas as Lord Kelvin and Lane thought, but rather a ball of plasma. It is a common mistake. In 1993 the band They Might be Giants released a record called *Why Does the Sun Shine?*, featuring a song of the same name (which was their cover version of a 1959 song). The strap-line for the song, and part of the lyrics, is the line 'the Sun is a mass of incandescent

gas', an almost exact quote from Lane. I was prepared to over-
look this persistent inaccuracy because I like any songs about the
Sun, but then They Might be Giants put it right when in 2009
they released a corrected song: 'Why Does the Sun Really
Shine? The Sun is a Miasma of Incandescent Plasma'.

The extreme temperature in the centre of the Sun means that
the atoms within it are torn apart into a plasma. The hydrogen
atoms have lost their single electron, the helium atoms have lost
both their electrons, lithium has lost its three, and so on through
the heavier elements. Once an atom has been stripped of its
electrons, only a naked atomic nucleus is left, and since hydrogen is the most abundant element in the Sun, the plasma is made
mostly of electrons and single protons. This is the environment
that the gamma ray photons shine into. So what happens to the
gamma ray photons?

It's a useful analogy now to think of the photons as waves on
an ocean. If you place a buoy on the ocean's surface, you can
watch it rise and fall as the waves ripple underneath it. If a wave
with a short wavelength moves the buoy, it will bob rapidly up
and down. Another wave, moving at the same speed as the first
but having a longer wavelength, will move the buoy up and
down less frequently.

If you have an electromagnetic buoy – a device to measure
the ripples in the magnetic and electric fields passing any one
point – you would see that the strength of those fields moves
continuously from high to low and back again. In fact, you do
have such a device in the form of your TV aerial. The electromagnetic waves passing through it make the free-drifting
electrons in metal bob around. This generates an electric current
that is then converted into images and sound. The important
point here is that the electrons act like a buoy for electromagnetic
waves. With this in mind, let's head back into the Sun.

As the electromagnetic wave of the gamma rays ripples out

from the core it encounters a fleet of liberated electrons. And, just like a buoy on the ocean, the electrons start to bob up and down. Each time this happens, energy has been transferred from an electromagnetic wave to an electron. The photon is actually 'absorbed' by the electron during this process, disappearing and leaving behind only a very excited electron. But that electron cannot contain its excitement. As we have seen, an accelerating charged particle emits radiation, so once the electron starts moving it sends out radiation of its own and emits another gamma ray photon. Crucially, though, the direction of that new flash of gamma ray light is completely random.

Standing back and watching this you'd see a photon hit an electron and almost immediately go flying off in a random direction. Officially, though, it is a new photon that is produced, but what we care about is the energy, and the quantity of energy is the same even if the actual photon isn't. Imagine that I lend you a £50 note and that you pay me back in a few days. I won't get the same £50 note back. But that doesn't matter – I'll get the same amount of money and will still be in an excited state. It's the same for photons. Photons are fungible.

The collective effect is as if all the photons were being constantly scattered in random directions by the electrons. This scattering is known as 'Thomson scattering' and is named after J. J. Thomson, the British physicist who discovered the electron in 1897, during his research to understand the workings of the atom. Each photon is being scattered, time after time.

In between each encounter with an electron, any one photon will be moving at the speed of light, but on average each photon only manages to travel about 1 millimetre in the radiation zone before hitting an electron and being scattered again. Not only is that 1 millimetre almost inconsequential compared to the overall thickness of the radiation zone it has to traverse (which stretches for over 400,000 kilometres above the core), but also it may not even be a millimetre in the correct direction. Each scattering event is as likely to send the photon back towards the centre of the Sun as it is to send it up towards the surface. The photon is on a random walk and this is why it takes so long for it to reach the surface. Each photon can expect to make around ten million billion billion steps before it randomly happens to reach the surface and get out. And this number of steps takes 170,000 years.

Photons find the Sun's plasma, with all its electrons, very difficult to get through. Their plight reminds me of one childhood memory that for some reason is still overwhelmingly vivid. It's night and I am in the car with my mother, who is driving, and the night is so foggy that we are practically hanging out of the car windows, straining our eyes to see what lies ahead. Even though the headlights are on, the photons barely penetrate into the haze. Instead, the car lights and the fog create a bright glaring wall of white immediately in front of us, as much of the light is scattered right back at us. It is a memory that flashes back to me whenever I feel claustrophobic. And this is what an anthropomorphized photon would be feeling: the sense of being boxed in, of its near-futile progress.

To make things even worse, the photons have to put up with the other by-product of the fusion occurring at the Sun's core: neutrinos. Each set of reactions that produces one helium nucleus releases two neutrinos as well as the two photons. But the neutrinos do not care about plasma. They race through it as it if weren't there, like car headlights through still, empty air on a summer's night. The neutrinos go from the core to the surface of the Sun in just two seconds! Imagine an exhausted photon, bounced around a millimetre at a time for hundreds of thousands of years, near the surface and probably only a few tens of thousands of years of random scattering away from making it out, when suddenly a neutrino less than two seconds old races by and doesn't even acknowledge its existence!

Even passing the surface is a non-event for a neutrino. It carries on at the speed of light regardless of what is going on around it. Eight minutes and 20 seconds later it reaches the Earth and pays us equally little heed. Hold up a finger right now and look at your fingernail – around 100 billion neutrinos are flying straight through your fingertip every second as if it weren't there. Only on very rare occasions will a neutrino interact with an atom, which is our only way to glimpse them. Physicists monitor thousands of tonnes of water in vast underground tanks just to spot a handful of neutrino–water interactions.

Hitching a ride

Things actually get slightly better for the photons when they reach the final 18 per cent of the distance out of the Sun and hit the convection zone. The pressure and temperature of the plasma have gradually been decreasing as the photons move further out. The centre of the Sun is a blistering 15 million Kelvin, but not all of the Sun is at that temperature: energy is constantly being lost from the surface of the Sun out into space. This means there is a temperature gradient from the centre of the Sun outwards: in the radiation zone the temperature drops to 7.5 million Kelvin and at the beginning of the convection zone it is a mere 2 million Kelvin – just as the further you sit away from a campfire in the cool night air, the more your temperature drops.

It's not just the relatively low temperature that is important in the convection zone though: it is the rate at which the plasma continues to cool as you go outwards towards the visible surface of the Sun. The steady drop-off in temperature in the radiation zone now becomes much more rapid, and this has an important consequence for both the plasma and the photons.

Very importantly, the photons have a much harder time getting through and the plasma starts convecting. Any rising bubbles of plasma that are hotter than their surroundings start to rise. But as they move higher, the rapid temperature drop in the plasma around them means that they have a good chance of still being hotter than their surroundings and they continue to rise. This is the cause of the thin convection zone which gives the boiling, mottled appearance to the Sun's surface. But what happens to the photons in this zone?

The temperature of the plasma has dropped enough for some of the particles, like carbon and nitrogen, to hold on to some of their electrons. Once this happens, the photons encounter not a plasma of just electrons and naked nuclei, but a plasma that also contains ions: 'ion' is simply the word for an atom that is electrically charged because it has managed to gain or lose some electrons, and doesn't have the right amount to be electrically balanced. Because the nuclei are only just able to start holding on to some electrons, they are still a long way off having all the electrons they need.

The problem for the photons here is that, unlike a free-floating electron, an electron attached to a nucleus has the ability to absorb a photon and it doesn't necessarily have to re-emit it. These are electrons with an eye on the future. If you loan money to these electrons you won't necessarily get it back. They have a very negative outlook on life. The reason why they can hold on to a photon comes down to their relationship with the nucleus.

As a solar physicist, I find the image of an atom being like a miniature Solar System, with the electrons racing around the nucleus as the planets orbit the Sun, very appealing. This model of atoms was developed by a Danish physicist, Niels Bohr, and a scientist from New Zealand called Ernest Rutherford, and is very sadly not correct. An Austrian scientist, Erwin Schrödinger, showed that the electrons move in much more complicated ways:

they don't sit in a neat circular orbits but rather spread across more than one location, actually more like a wave than a discrete particle. In fact, it becomes impossible to say where the electron actually is. Schrödinger's description was of a cloud around the nucleus in which he could assign the probability of the electron's position. This cloud has become known as an 'orbital' – a nod to the naming in the old Solar System-like model.

Thankfully, the Solar System analogy works perfectly if you replace the word 'orbitals' with 'orbits'. In the Solar System, not all orbits are equal and it takes different amounts of energy for planets and other bodies to remain orbiting at different distances. At this moment the Moon is losing energy because of its gravitational interaction with the Earth, so its orbit is changing: the Moon drifts away from the Earth by about 3.7 centimetres per year. Likewise, different orbitals around the nucleus require the electron to have different amounts of energy.

Crucially, unlike gravitational orbits that can have any amount of energy, the electrostatic orbitals can only take on set amounts of energy and nothing in between. It is as if, instead of gradually drifting away, the Moon suddenly jumped a few metres each time its energy matched an allowed orbit. And different kinds of nuclei, made of different combinations of protons and neutrons, allow for different energy levels that an electron can be in.

There are very specific energy levels associated with each type of nucleus and this means that the electron isn't free to choose the energy it can have. It must occupy one of the predetermined energy levels. So a photon will only be absorbed if its energy exactly matches the amount required to shift the electron between energy levels or if it can eject an electron completely. If there is no match the photon will simply pass by.

And if the photon is absorbed when its energy matches, the ion may not release it again straight away. On top of that, the ion is also in motion: it is riding the convection currents up and down the convection layer. A photon can be absorbed by an ion, ride along with it up the convection zone and then be re-emitted closer to the surface. With an average speed of 1 kilometre per second, hot pockets of plasma move across the 125,000 kilometres of the convection zone in a couple of days. It takes far less time for photons to cross the convection zone than to cross the radiation zone, as they are able to hitch a ride for the final stages of this journey. Convection takes over from radiation as the main way to transport energy.

The very last step of the journey is the photosphere – the sphere of light – where the plasma finally breaks its hold on the photons and they escape into the Solar System. It's an incredible layer and is only 500 kilometres thick – less than one thousandth of the solar radius. Over a distance of 500 kilometres the plasma goes from being an opaque barrier to the photons to being transparent. Such a rapid transition gives an otherwise tenuous ball of gas, which has no surface, a very sharp edge indeed when we view it. No wonder that over the scientific generations there have been some who were convinced that the Sun is a solid object. And without this sharp visible edge, we wouldn't be able to ascribe a particular size or shape to the Sun at all. It would look like a hazy amorphous blob.

The clear and sharp edge, which looks so solid, belies the true density of the plasma, which is astonishingly thin. At the base of the photosphere, the plasma is 10,000 times less dense than the air around us. The temperature has also plummeted; the photosphere goes from only 6500 Kelvin to an almost chilly (relatively speaking) 4900 Kelvin at the outermost edge. Even though the plasma continues to extend well beyond this – more on this later – these are the temperatures and densities at which the photons can suddenly pass through the material unimpeded, finally make some distance and go streaming out into the Solar System.

But they have not survived the journey unscarred. Each of those ten million billion billion scattering events had a slight impact on the photons. Unperceivable individually, only across so many collisions can the slight amount of energy that is lost each time start to accumulate. During the long radiation zone journey the gamma ray photons lose a little energy and become X-ray photons instead, and then, by the time they encounter the convection zone, they have lost enough energy to be ultra-violet photons. During the convection-zone journey the gradual leaking of energy brings the photons down into the range, by the time they escape the photosphere, visible to our eyes, giving us the sunlight we all know and love – with two slight problems. All through their journey, the photons have been pinged around by individual electrons and eventually individual ions. This means they should form a spectrum with thin discrete spectral lines – which is not what we see coming from the Sun. And the extremely low plasma density at the bottom of the photosphere should mean that the photons can easily fly through it – which is not what is happening. Photons are absorbed and emitted across a broad range of wavelengths covering the visible part of the spectrum and this is a puzzle when set against the description of the thin photosphere that we have developed. There is one last character-changing adventure that the photons experience in the thin photosphere.

The negative hydrogen particle

Firstly, the behaviour of electrons in orbits explains the emission spectra we saw in Chapter 3. When the electrons in an ion (or atom) lose energy and emit a photon, they can only move between certain set energy levels, and so the released energy can only have specific values. The energy of the photon determines its wavelength/frequency and this is why different elements release light of very specific colours. It is also why they can absorb only specific colours of light.

To come back to our financial analogy, consider bank A, which only accepts deposits or withdrawals of £4.10, £4.34, £4.86 and £6.56,* and bank B, which only accepts £5.16, £5.17 and £5.18.† Just knowing these numbers allows us to recognize which bank money is going to or coming from by looking at the amounts being transferred.

The problem is that we do not see an emission spectrum coming from the Sun, but instead a continuous spectrum that has an absorption spectrum removed from it. Half of this makes sense. The Sun's atmosphere continues out past the photosphere and there are hydrogen, helium, lithium, iron and whatnot floating about in it. If a continuous spectrum were somehow produced by the plasma in the photosphere, it would have to pass out through these elements in the overlying plasma, which explains the absorption lines.

It turns out that the continuous spectrum was coming from a very shadowy figure, a particle no one expected: a strange and exotic variation of hydrogen – the so-called 'negative hydrogen ion'.

A hydrogen ion with no electrons is a lone proton floating about by itself. This is the case at the base of the photosphere, where the temperature of 6500 Kelvin is enough to stop any of these loose protons from hanging on to an electron. But 500 kilometres later, the temperature has dropped sharply to around 4900 K. Now the temperature becomes cool enough for the

---

* The visible spectrum of light from hydrogen contains spectral lines at four wavelengths, 410.2, 434.1, 486.1 and 656.3 nanometres. These are known as the Balmer lines.

† There are three prominent absorption lines in the solar spectrum from neutral magnesium at the wavelengths 516.7, 517.3 and 518.4 nanometres.

hydrogen nuclei to capture their electron and form a neutral hydrogen atom again. Meanwhile, the electrons orphaned by much larger atoms are still present, as they find it even harder to hold on to their outermost electrons. As these orphaned electrons move around, they collide with the newly formed hydrogen atoms around them and something new is formed.

The simplicity of hydrogen, which has only one electron in orbit around one proton, means that the electron is not able to provide a good shield to the positive charge of the nucleus. Orphaned electrons in the vicinity feel this weak electrostatic charge and occasionally are captured to give a neutral hydrogen atom an extra electron. The two electrons provide two units of negative electric charge and this outbalances the single unit of positive charge in the nucleus, and so we have the negative hydrogen ion.

That loose bonus electron rattling around is the secret of the negative hydrogen's continuous spectrum. The negative hydrogen ion is able to lose its extra electron through the absorption of a photon across a wide range of energies or wavelengths in the ultraviolet, visible and infrared ranges. The key is that instead of having to bump the electron into a new orbital, it knocks it off completely. This is a much more flexible process. When a new bonus electron is captured by a hydrogen atom, the process is reversed, leading to the production of a continuous spectrum of light across these wavelengths.

The negative hydrogen ion is responsible for 95 per cent of the light emitted by the Sun. It is easy for the hydrogen atom to gain or lose an extra electron. And when a photon is emitted it can be at a range of frequencies across the ultraviolet, visible and infrared parts of the spectrum. And this matches exactly what we observe.

You may find it bewildering to know that there is only one negative hydrogen ion to every one hundred million hydrogen atoms in the photosphere. Yet this is enough to control the flow of the light. That's like saying in a country the size of the US with 300 million people only three of them control the entire flow of money through the economy. The Sun seems to be driven by some extremely unlikely processes: from the creation of light itself, which relies on two protons fusing to start the chain – something that only happens once in a hundred million collisions – to the release of that energy at the photosphere, which relies on a particle that is one in a hundred million. But, of course, the sheer scale of the Sun makes these things possible.

We did not understand the role of the negative hydrogen ion in facilitating the escape of photons from the photosphere until 1938, just one year before Bethe's work on explaining the rate at which nuclear processes are happening to generate the Sun's energy. The negative hydrogen ion was deemed theoretically able to exist in the 1920s, and at the time the energies involved in a hydrogen atom capturing and losing an extra electron were being investigated. Progress was slow, though, and the early results were inconclusive and sometimes even contradictory. It took the new field of quantum mechanics to provide a way to solve these uncertainties – again, the work of people like Schrödinger and his theory of describing particles as waves provided a mechanism by which to investigate them, and in 1929 and 1930 it was shown that a negative hydrogen ion could be formed and that it could be stable.

With the theoretical foundations laid, it was Rupert Wildt, at Princeton University Observatory, who suggested that the negative hydrogen ion might be responsible for controlling the flow of photons through the photosphere and hence the sunlight we receive. The founder of space science at University College London (UCL), Sir Harrie Massey, was also involved in this work. His contribution on the details of the absorption of photons by the negative hydrogen ion was key for gaining acceptance of the role of this particle in governing how light travels through the photosphere of the Sun, and, indeed, other similar stars. The Sun's visible output was finally understood. The negative hydrogen ion is the particle that absorbs and emits photons in the photosphere, despite there being so very few of them.

What we now call the 'standard model' of the Sun was completed in 1957, after almost 100 years of development from the time of the early work of Lane. Which is not to say the story is over. The standard model of 1957 was lacking in one major aspect: it had no way to explain the presence of sunspots – dark ominous regions that regularly appear on the surface of the Sun.

5. Sunspots

As the Sun began to rise over London on 8 December 1610, there was frost on the ground and a thick mist was in the air. That morning a British astronomer and mathematician, Thomas Harriot, was poised and ready to do something that no one had done before: look at the Sun through a telescope. The telescope had been invented in Holland and was publicly known about by 1608. But the telescope had been used only for distant terrestrial objects, useful for the Dutch, no doubt, who were at war with Spain at that time.

Harriot's problem was that the light from the Sun is too bright to look at directly at the best of times – using a telescope to gather even more sunlight was going to cause serious damage to anyone who put an eye up to the telescope. His solution was delightfully low-tech. But dangerously so. He waited for a very misty morning and looked at the Sun first thing, when it was just a few degrees above the horizon, so its light shone through as much mist as possible. He was using the mist as a kind of filter to dim the Sun's light. This did reduce its brightness a lot, but not enough to be able to stare at it through the telescope. So Harriot flipped frequently between his left eye and his right so the light couldn't dazzle, and cause too much damage to, either one of them. This method is not recommended!

We still have the notes he made about what he saw, and accompanying his writing is a drawing of how the Sun appeared to him that morning. It shows that on the circular disc of the Sun were three black spots: sunspots. We now know that the Sun is frequently covered in these black spots of changing size and position. In fact, observations across the centuries after Harriot have made us realize that sunspots are invaluable in our study of the Sun. They have a central importance because they allow us to scratch the surface of our star and reveal that it is not a perfect luminous orb and that it changes over time – its appearance varies.

In fact, who it was that first viewed the Sun with a telescope and saw sunspots is still debated. Galileo is the name that normally springs to mind but it seems he did not make any record of sunspots until April 1612 – two years after Harriot's drawing. Galileo's story is remembered because of the vast and important range of astronomical observations that he made but also because his story is so captivating. His ideas about a Sun-centred Universe, around which the planets orbited, were at direct odds with the Roman Catholic Church, which still held the Aristotelian view that placed the Earth at the centre. The Church also proclaimed that the Sun was perfect, immaculate and, therefore, blemish-free. It was a discovery counter to doctrine and this culminated in Galileo's trial in 1633 by the Roman Inquisition, who found him to support the idea of a Sun-centred Universe and ordered that he remain under house arrest for the rest of his life.

Slightly ahead of Galileo was Johannes Fabricius, in Germany, who first observed the spots in 1611 and made sure he told everyone about them, publishing what he saw in a pamphlet. Thomas Harriot in England, who drew spots in December 1610, would have predated Fabricius but Harriot never published his work. Instead, his manuscripts remained hidden for many years after his death in 1621, only coming to light in the late 1700s. Some people (myself included!) now consider him the first true observer of sunspots through a telescope, while others still back Fabricius. Interestingly though, Harriot's 8 December 1610 diagram clearly shows sunspots, but he does not mention them directly in his text – perhaps because he did not have the words to describe what he saw or because he couldn't realize their significance at the time he made that first drawing.

Whoever the first observer was, we know for sure that a German called Christoph Scheiner was the first person to take sunspots seriously. Starting in 1611, Scheiner made a comprehensive and dedicated survey of sunspots over the following years, publishing them in the pioneering tome *Rosa Ursina, sive Sol* in 1630 (confusingly, much of his work was published under a pseudonym). This was a significant record of his careful study of sunspots' shapes, lifetimes and motions that became the definitive work in this area for over 100 years. Scheiner also contributed to the development of telescopes to view the Sun, using coloured lenses instead of clear ones so that the Sun's brightness could be diminished. He called this modified instrument a 'heliotropii telioscopici', which is normally translated as a 'helioscope'. This method is the recommended one! Today many people enjoy back-garden solar astronomy by putting a specialist solar filter (much darker than sunglasses or even welding masks) over their telescope before pointing it at the Sun.

There are some observations of sunspots which predate telescopes. Naked-eye observations, although very dangerous, can be made when sunlight is diminished by fog or cloud. Given the curiosity of humans, with no other way to do it safely, people have been eyeballing the Sun for millennia. Evidence that the Sun was looked at directly can be found in records in Korea, Japan, Vietnam and, most famously, Ancient China which stretch at least as far back as 165 BC. The persistence of Chinese astronomers is to be admired as the records show that on average only one sunspot was seen every decade. What the ancients saw, though, gave crucial clues about the appearance of the photosphere and their records provide a valuable dataset that has allowed us to investigate how the Sun changes over time and is still used today, over 2000 years later.

The Ancient Greeks also saw spots. Theophrastus of Athens, a student of Aristotle, reported them in his work *De Signis Tempestatum* and discussed the occurrence of sunspots in relation to the weather, even using them to make weather forecasts – one of his less successful ventures. But he made up for it with pivotal works on biology, ethics and the most definitive book on stones ever written.

With the power of hindsight it even seems that there were some, albeit rare, European records of observations of sunspots before the development of the telescope. In 1128 an English monk, John of Worcester, made a drawing of the Sun with two black spheres against its disc, and in 1590 Henry Hudson was sailing on the ship *Richard of Arundel* just off the coast of West Africa, when he recorded spots on the Sun as he looked at it at sunset. But the spots weren't recognized at the time as being sunspots or thought to be important. Even Johannes Kepler, whose work was central to the development of European astronomy, saw a black spot in 1607 when he projected an image of the Sun in Prague. Missing an opportunity to present the discovery of a new feature, he misinterpreted the black spot as being the planet Mercury transiting in front of the Sun.

The sunspots viewed through telescopes began to show certain consistencies which revealed information about the Sun that could not be known before. Galileo had mainly been using a projection technique to make fantastic observations of the Sun. Instead of looking down his telescope at it, Galileo let the light flood out of the eyepiece and project the image onto a piece of paper (much safer than projecting it on your retina!). A student of his had discovered the method and it meant that the Sun could be observed even at midday when it is at its highest point and the sky is more likely to be clear. And smaller spots could be made out using this technique than by looking directly through the telescope.

Galileo's drawings showed that the sunspots consistently moved across the disc of the Sun from left to right. And when the spots were at the centre of the Sun they looked round, like circular plates viewed from above, but near the edge of the Sun the spots narrowed, like plates looked at side on, an effect known as 'foreshortening'. This was the first conclusive evidence that the Sun was a sphere and that it was spinning. But it also showed that the sunspots were a feature of the solar surface, and not objects between the Sun and us. But what were they?

The shape of the spots

Scientists made much progress in describing what sunspots looked like before they had any accurate interpretations of what those observations actually represented. Scheiner had shown that sunspots have a central black, round region that we call the 'umbra', which is surrounded by a slightly lighter region, named the 'penumbra'. Sunspots formed by gradually increasing in size and decayed by reducing in size and fragmenting into smaller spots. And once a sunspot decayed, the area it had previously occupied looked exactly the same as the surrounding solar surface – as if nothing had ever been there. The spots tended to appear in two horizontal bands roughly 30–40 degrees above and below the equator and, most confusingly, bar a few lone cases, they always seemed to hang out in pairs.

The first insight into what sunspots were came in 1769 when an extremely large spot appeared and started to make its way across the Sun. Just as it does today, the appearance of the large sunspot sparked interest and excitement. Such sunspots are relatively rare – I will still get a flurry of emails from friends and colleagues when a new mammoth sunspot appears to make sure I don't miss it. Knowledge of the extraordinary 1769 sunspot spread amongst astronomers and quickly reached the chair of astronomy at Glasgow University, Alexander Wilson, after a friend in London notified him. As Professor of Practical Astronomy, Wilson was keen to see the spot for himself and monitor it for any change in size.

On 22 November 1769 he turned his telescope to the spot and viewed it, magnified 112 times. It must have looked magnificent. When he first saw the sunspot the penumbra was equally broad all the way around the umbra. What he saw during his observations on subsequent days intrigued him. As the sunspot moved away from the centre of the Sun and towards the edge, Wilson saw that the penumbra on the side of the spot *furthest* from the edge of the Sun had contracted in width and was narrower than the penumbra on the other side of the spot. This observation was remarkable because it was exactly the opposite of the foreshortening that Wilson expected: the penumbra on the side *closest* to the Sun's edge should contract the most because this side of the spot turns away from us first.

Wilson suspected that this sunspot represented a physical depression in the surface of the Sun. Sunspots might be sunholes. He theorized that the umbra was the bottom of a deep excavation and the penumbra the downward-sloping sides — more like a bowl than a plate on the solar surface. Looking into this depression meant that you were able to see the sloping side closest to the edge of the Sun because it turned to face you. In contrast, the side closest to the equator becomes hidden from view. This was an elegantly simple interpretation of the observation and Wilson wanted to test it to see if it held up at other times and when sunspots were close to the other edge of the Sun. So he waited, hoping that the large spot would live long enough to survive its two-week journey around the far side and make another appearance on the face of the Sun. And on 11 December the spot returned.

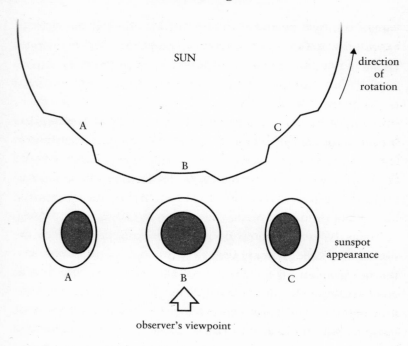

5.1 The uneven contraction of the penumbra on the sides of sunspots when they are near to the limb led to sunspots being interpreted as depressions in the photosphere.

As it peaked around the left side of the Sun only three sides of the penumbra could be seen around the sunspot's dark centre. The now centre-ward side of the penumbra, which had been visible two weeks earlier, was missing. The next day the missing penumbra had appeared but it was very narrow. These observations were exactly as Wilson had expected, based on his idea that sunspots are depressions in the surface. After 160 years of telescope observations a testable theory about the nature of sunspots had finally been proposed, tested and backed up by further observations.

Wilson then went on to estimate the depth of the umbra

using his measurement of the breadth of the penumbra at full extent, the mathematics of triangles and the radius of the Sun. He proposed that they were at least 6000 kilometres deep, just a small notch compared to the radius of the Sun though, leading him to conclude that the Sun was a solid, dark sphere surrounded by a thin luminous substance – the idea being that sunspots are regions where you can see through the luminous layer to the sphere within. This view of the construction of the Sun prevailed for many decades, with a similar idea being proposed by Herschel almost three decades later. It was a while before Lane and Lord Kelvin were able to suggest otherwise.

Today, Wilson's depression is understood as being a depression in the *visible* photosphere of the Sun. For some reason, in a sunspot the photons are able to escape the plasma sooner than in other regions. The plasma must be cooling down more in sunspot regions than it does elsewhere. The relative darkness of sunspots indicates that they contain plasma that is around 2000 Kelvin cooler than the surrounding photosphere and the transparency of this relatively cool plasma, controlled by the negative hydrogen ion, is heavily affected by its temperature. In the sunspot, the plasma is more transparent and we can see further into the Sun – hence the apparent depression.

The magnetic Sun

Understanding what caused sunspots fell to an American astronomer, George Hale, who was born into a wealthy family in Chicago in 1868. He grew up in an era when the developments that underpin modern science were being born and he was an insatiable scientist himself. Even as a young boy his relentless drive led him to set up his own private observatory, funded by his wealthy father. The observatory was well equipped and was a first-rate facility in which Hale could do astrophysical research. Hale had a passion not only for science, but also for engineering. He could design and build the instruments that he needed for his investigations.

And it was an instrument of Hale's making which gave us the required insight into sunspots, an instrument that I use to this day to continue this study of sunspot regions. Called the 'spectroheliograph', Hale's equipment was a steam punk-esque creation made out of glass and mirrors, wood and metal. I use the modern equivalents today, which need electronics and are built into spacecraft, but the basic principle of how they look at the Sun is exactly the same.

When Hale was young, the Sun was typically studied using telescopes that made an image using the entirety of the visible radiation, just as our eyes do. Photographic plates replaced the astronomer's eye so that a permanent record of the Sun and its spots could be created – this kind of instrument is known as a 'heliograph'. It was effectively pointing a camera with a zoom lens at the Sun: all the light coming from the Sun was allowed to reach the photographic plate.

The developments in spectroscopy had led to the use of a second instrument, where the Sun's light was split into its spectrum so that the individual spectral lines could be analysed. These instruments are called 'spectroscopes' and they are vital because they allow the chemical composition of the plasma to be probed without the need to visit the Sun directly. But they are not fundamentally different to Newton letting a beam of sunlight hit a glass prism and cast a spectrum on the wall.

Hale devised a third way to view the Sun. The spectroheliograph is a photographic telescope combined with a selective spectroscope. A telescope is still used to magnify the Sun and project its image onto a photographic plate, but along the way all frequencies of light but one are removed.

The business end of the spectroheliograph pointing at the Sun was a thin slit. It only allowed the light from one vertical line across the Sun's surface to enter the device. You can imagine the reverse of this set-up, looking out through the slit towards the Sun: you won't see the full Sun – just a narrow strip from top to bottom.

Once inside the telescope this light was cast onto a prism, splitting the light into a full spectrum (or a grating could be used instead of a prism to produce the same result). This spectrum was cast onto an obstructing panel. This stopped all of the light from going any further, except for where one tiny slit in the panel allowed the light from just one spectral line to pass through onto a photographic plate behind and expose a vertical stripe.

So far this probably doesn't sound like a particularly useful device, or even significantly different from a spectroscope. But when the entrance slit is gradually moved across the entire Sun and the photographic plate is moved in unison, you progressively expose an image of the entire Sun in only one wavelength of light. Importantly, that one narrow wavelength range you allow through corresponds to a specific movement between orbitals in one given element. Using a spectroheliograph you can take a photo of the Sun and only see the magnesium emission, completely removing the dazzling light from all other elements. Or you can choose any element you please (as long as it is emitting light).

Building this kit was no easy task – any vibrations from the moving telescope will blur the photograph. But by 1891 Hale, who was a final-year undergraduate student at the Massachusetts Institute of Technology at the time, had successfully crafted one.* Hale went to work at the University of Chicago and from there founded the Mount Wilson Observatory, about a fifty-kilometre journey north-east of Los Angeles in the US. He

---

* Around the same time a British astronomer, John Evershed, and a French astronomer, Henri Deslandres, independently also devised similar instruments.

resigned from the University of Chicago to work full time at Mount Wilson in 1905.

Hale had almost exclusive use of the spectroheliograph at Mount Wilson and he used it to study the Sun's lower atmosphere, first of all using a spectral line formed by calcium ions that have lost one of their electrons. This ion is formed in the plasma in the photosphere and the region of plasma just above it. By imaging the Sun in this one specific frequency only produced by calcium in certain situations, Hale could selectively take a photo of just the one aspect of the Sun. He was imaging the Sun in ways that would transform solar research.

Hale also used light emitted by hydrogen atoms. After all, this is the most abundant element in the Sun and there are several spectral lines that hydrogen produces. In the visible part of the spectrum are four prominent lines, 'hydrogen alpha', 'hydrogen beta', 'hydrogen gamma' and 'hydrogen delta'. These form a sequence by wavelength, going from hydrogen alpha, which has the longest wavelength (in the red part of the spectrum), to hydrogen delta, which has the shortest wavelength (and is in the violet part of the spectrum), each one formed by electrons moving between different energy levels (orbitals) leading to the emission of light in the visible part of the spectrum. Hydrogen beta is in the blue part of the spectrum and hydrogen gamma at the blue/violet interface.

Out of these hydrogen spectral lines it was those in the blue and violet parts of the spectrum that were first selected for use in the spectroheliograph, and for purely practical reasons. Images of the Sun were being recorded on photographic plates, and in the early days of using this technique no commercial plates were available that were sufficiently sensitive to red light. But Hale was curious to know what the Sun would look like in the light of hydrogen alpha so he was in luck when, at the end of 1907, developments in the chemical preparation of photographic plates made it possible for images to be created in light at the red end of the spectrum.

The hydrogen alpha line has a wavelength of 656.3 billionths of a metre and is formed when an electron falls from the third to the second energy level in a hydrogen atom, most likely when a free electron recombines with a hydrogen nucleus and cascades inwards. The light we see at this wavelength is coming from a height of around 5000 kilometres above the photosphere, from glowing plasma directly above sunspots. In my research into the conditions above sunspots I still use hydrogen alpha images.

What Hale saw at Mount Wilson on 28 March 1908 changed solar physics for ever. He loaded in a new-fangled photographic plate sensitive to red light, set the spectroheliograph to the hydrogen alpha wavelength and scanned it across the Sun. The image it produced revolutionized our understanding of the Sun. Although the image was faint, it showed signs of structure above sunspots. And by 30 April Hale had mastered taking images in this new wavelength and saw magnificent whirling vortices of plasma. There was structure and detail in the plasma above sunspots that was clear to see. But the image taken on 28 March 1908 is, for me, the image that changed the direction of solar physics from counting and recording the positions of sunspots to understanding the physics of them.

In hindsight, there had been hints before of the whirling structure seen in the light of hydrogen alpha. But these images were so clear that they triggered Hale's interest and got him thinking. As Hale considered these whirling structures he began to see connections with other developments in physics.

In 1876 experiments that would eventually lead to our understanding of electricity and magnetism were just taking shape. At the time people were marvelling that if an electrically charged rubber disc was set spinning, it produced a deflection of a compass needle. They did not realize that in this movement of

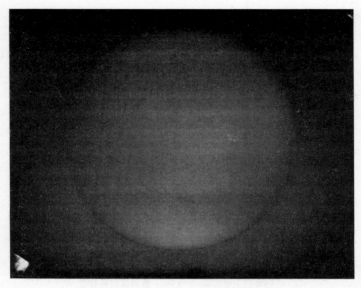

5.2 First image of the Sun taken in the light of hydrogen alpha (*Carnegie Observatories*).

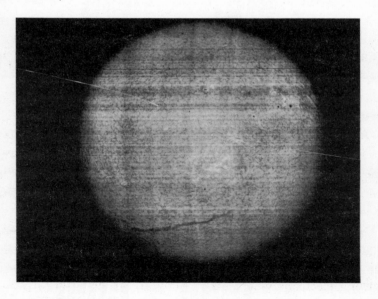

5.3 Image of the Sun in the light of hydrogen alpha, taken on 30 April 1908 at the Mount Wilson Observatory (*Carnegie Observatories*).

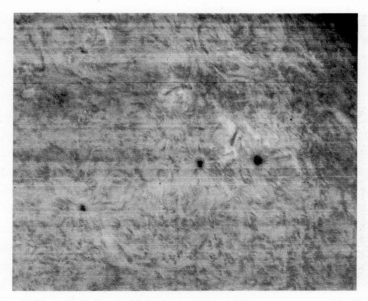

5.4 Zooming in on the sunspots shows the radial features around the spots that were called 'vortices' by Hale (*Carnegie Observatories*).

electrical charges causing a magnetic field they were opening a Pandora's box of modern physics. Only in 1897 did the scientist J. J. Thomson discover something he called the 'electron'. In 1908 there was still not complete agreement as to what atoms were or even if they existed.

Hale took these latest findings and speculative results and used them to explain what was going on above sunspots. He reasoned that the high temperature of the solar atmosphere would mean that the particles there are torn apart so that electrons would be separated from their parent atom and moving freely in the solar atmosphere. Hale saw the newly confirmed vortices, and the sunspots that they connected to, as regions of swirling electrons where magnetic fields are being generated. Which was a bold judgement; at that point the only magnetic field that was known to exist in the entire Universe was the Earth's. Was the Sun going to be the second? Hale knew just the way to find out.

The last piece of information that Hale drew upon was an 1896 experiment by a Dutch physicist, Pieter Zeeman, who had shown that a strong magnetic field could affect the light given off by a 'luminous vapour'. If a gas was in a strong magnetic field when it produced its emission spectrum, the magnetic field had an effect on the spectral lines. A weak magnetic field would cause the lines to get wider and a strong magnetic field would cause some of them to split entirely into two or more separate lines.

Hale wasn't the only one to realize the astronomical potential of this discovery. Indeed, Zeeman's paper, published in 1897, already speculated that this discovery might be used to detect cosmic magnetic fields, including any present in the Sun. The effect of a magnetic field on spectral lines is now known as the 'Zeeman effect'. Zeeman was awarded the Nobel Prize in 1902 jointly with Hendrik Lorentz, someone whom we will meet again shortly, for their work in this area.

Hale knew that if magnetic fields existed in sunspot regions, and if they were strong enough, they could be detected by telescopes on the Earth by means of a detailed study of spectral lines as Zeeman suggested. And Hale had both the scientific knowledge and the technical ability to test this. Sure enough, when the spectral lines of light produced in sunspot regions were carefully observed, they showed the broadening and splitting which is characteristic of the presence of strong magnetic fields. When Hale studied the spectral lines of light coming from the surface of the Sun where there were no sunspots, the spectral lines looked normal.

Hale had done it. Sunspots, which were often many times the size of the Earth, had been found to be the source of a magnetic field many thousands of times stronger than the magnetic field at the surface of the Earth. The paper that Hale published on this in 1908 (within three months of taking the 28 March image) is now regarded as the birth of modern solar physics. In fact Hale had already started thinking that sunspots might be regions of magnetic field in 1905. But his observation – direct proof that he was right – allowed the study of the sunspots to move beyond simply recording and classifying them, to beginning to understand their physical origin. From this point on we could finally start to comprehend not only what the Sun *is* but also *why* it behaves the way it does. And it gave the first evidence that magnetic fields existed beyond the Earth.

All Hale

It may seem like a failure of science that the true nature of sunspots wasn't understood until 1908 – almost three centuries after the first telescopic observations. In reality, though, understanding sunspots wasn't possible until spectroscopy had been developed, along with an understanding of how changes to spectral lines could be used to infer the presence of a magnetic field at the light source. But this discovery of the magnetic fields associated with sunspots suddenly explained all the mysterious observations.

This was why sunspots came in pairs: they are a pair of north and south magnetic poles. The names of 'north' and 'south' for magnetic poles are a throwback to Earth's magnetic field remaining (relatively) fixed from pole to pole. On the Sun, we tend to call the two magnetic ends of a pair of sunspots the 'positive' and 'negative' magnetic fields. This helps convey the sense that in the north polarity the magnetic field is pointing out of the Sun (positive) and in the south polarity it is pointing into the Sun (negative).

If you think back to the bar magnet diagrams you saw at school, with the field lines looping around from north pole to south pole, this is almost exactly the same as what is happening on the Sun. The magnetic field lines come straight up out of a positive-magnetic-field sunspot, loop around high above the photosphere and come back down through the other sunspot in the pair, the negative-magnetic-field sunspot. Instead of the iron filings used in the classroom to see this magnetic structure, in the Sun Hale was seeing plasma guided by the magnetic field (delightfully, containing iron ions).

In the first chapter I described a magnetic field as being a kind of field of influence, conceptualized using field lines that are like elastic bands that loop back around on themselves. This convention of field lines started with Michael Faraday, the great British scientist and popularizer of science, when he needed a visual image to help describe this invisible force. Faraday introduced the concept of 'lines' of magnetic force. These are purely imaginary but Faraday was a great communicator and this conceptual approach to magnetic fields is still invaluable today.

Faraday's concept of magnetic lines of force helps us visualize how this force varies across space, using lines that connect between the north pole and south pole of a magnet. And just as contours on a map indicate the steepness of a slope, magnetic field lines are more concentrated in regions where the force is strongest and more spread out where the force is weak. I think of them as elastic bands because, as we can already start to see from Hale's image, these field lines can become twisted and stretched. This was later to become more important in understanding the Sun than even Hale realized.

The discovery that sunspots are regions of intense magnetic field is also able to solve the riddle of why they are cooler and darker than the surrounding photosphere. Most of the plasma in the photosphere has a temperature that's close to 5700 Kelvin, whereas in the sunspots the temperature can drop to 3700 Kelvin. This is because the strong magnetic field disrupts

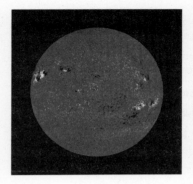 

5.5 and 5.6 Modern magnetic maps of the Sun. The positive (north) polarity
is always colour-coded in white and the negative (south) in black. The image
on the left is the observation, the image on the right has had the magnetic
field lines added, along with the numbers assigned to the sunspot groups
(*NASA/SDO, HMI science team and LMSAL*).

the convection that resupplies hot plasma to the photosphere.
Normally as plasma loses its energy into space and cools, it
would sink, only to be replaced with hot plasma, rising up
fresh from within the convection zone. The intense magnetic
fields trap plasma though, stopping hot plasma from coming in
and letting it continue cooling.

The picture which has since emerged from Hale's work is that
sunspots do not produce this magnetic field – it is the magnetic
field itself which produces the sunspots. A sunspot is just the
visual manifestation of what happens when a massive arch of
magnetic field penetrates the photosphere in two places. When
we look at a sunspot it is like taking a horizontal slice through
that intense bundle of magnetic field.

Magnetic fields cannot ever simply end, either; they always
loop back to where they started. The fields are three-dimensional,
which means there is much more to them than the cross-section
that forms a sunspot. The bundle of intense magnetic field that
comes out of a positive sunspot is like a 'tube' of intense magnetic

field that extends high above the photosphere. It then bends over to go back down through the photosphere in a negative sunspot, but the same thing must be happening below the surface of the Sun. Somewhere deep in the Sun the magnetic field must complete its loop. But it was a few decades after Hale before we could follow the magnetic fields on their journey back into the Sun.

As always, no physics theory is perfect straight away. There were a few observations that did not completely match this magnetic description of sunspots. Occasionally there would be a lone sunspot which did not seem to have a partner. These 'alpha' sunspots are a problem as a magnetic field never exists with only one pole. We now know that the magnetic field of the other pole is so spread out it is too weak to interrupt the convective flows in those places and no visible change in the photosphere temperature occurs. Slightly harder to explain is how the plasma in a sunspot cools to 4500 Kelvin but then somehow maintains that temperature without cooling any further, as we would expect it to. It's still not well understood how this is happening – some hot plasma must be being delivered still. Even 100 years on from Hale's discovery we still have questions about sunspots that need to be answered.

Hale died in 1938 but he left a huge legacy to solar physics and astrophysics in general. He saw the Sun as a typical star and never became exclusively a solar physicist; there were the observatories that he founded (four of them in total), the young scientists he supported at Mount Wilson, who include Edwin Hubble and Harlow Shapley, the journal that he founded and edited for most of his life (the *Astrophysical Journal*), which is one of the premier journals for the research community today, and the work he did in building and supporting solar physics and astrophysics. I am hard pushed to think of someone who had a greater impact across so many areas.

6. The Spinning Sun: The Day the Sun Fought Back

This chapter is about how fast the Sun is spinning. If this were a book about the Earth, the chapter on how long it takes to rotate would be a very short one indeed. Actually, it would just be a page that says: 'Twenty-four hours'. Or maybe '86,400 seconds'. There could be some confusion about whether we measure the rotation relative to the Sun or the stars (a difference of less than four minutes), and the rotation does vary ever so slightly (the day has increased by about 1.7 milliseconds over the past 100 years). But these are minor details. For the most part, because the outer layer of the Earth is a solid lump of rock, it all rotates in unison.

The Sun is far more complicated. As a giant ball of plasma it is a fluid object and does not all have to rotate in unison: different parts of the Sun can be spinning at different speeds. In addition, it can be hard to observe how fast bits of it are moving because, with the exception of sunspots, it just looks like a bright blank disc to us. But, as always, scientists have come up with some pretty ingenious ways to find the answers they are after, but there remains a question: why do we care?

I am a career scientist myself, so like generations of people before me the innate human curiosity to understand the world around us is more than motivation enough. All the people I've covered in this book have striven to understand the Universe around them to sate their inquisitive minds. Sometimes it was out of curiosity about the Sun, but mostly it was just out of a curiosity that then found an application in understanding the

Sun. But in 1859 the Sun gave a motivation to hurry along the investigation for pragmatic reasons: it took a shot at us. It started innocently enough. During the night of 1 September 1859 there was an intense display of the aurorae that erupted all over the globe. Normally a beautiful but rare natural occurrence unless you live at high latitudes, the aurorae were seen by the masses that night because they moved from their usual position around the poles and, in the northern hemisphere, reached as far south as Florida. Some people panicked, thinking that great fires had started. But most people just gathered on street corners to admire the spectacle.

Telegraph operators were not having such a great time though. The telegraph was the internet of the nineteenth century and was vital for sending information all around the globe. Suddenly the operators were struggling to send their messages and many lines had to be closed. Some telegraph offices caught fire as the equipment started to spark uncontrollably. Others disconnected their equipment from its batteries in an attempt to shut it down but found that the lines continued to operate with no battery attached at all.

And, at the same time, the magnetic observatories at Kew and Greenwich in London recorded something astonishing. They had magnetometers monitoring the Earth's magnetic field and they suddenly detected large changes. At one point the recordings even went off the scale. The Earth's magnetic field had suddenly and drastically changed. Then, over a number of days, things gradually returned to normal: the telegraph lines started working normally again and the aurorae receded back to their oval round the poles.

There had been some warning of this spectacle though. Seventeen and a half hours earlier a British brewer and amateur astronomer, Richard Carrington, had been projecting an image of the Sun onto a screen so that he could study its appearance that day. He had developed a very good method of filtering and projecting the Sun in order to measure and record the positions of sunspots. But going about his routine on that day he suddenly became, in his own words, an 'unprepared witness of a very different affair' when, within a large sunspot group, two patches of intense white light flared into sight. At the same moment, thirty kilometres to the north at Kew, the magnetometers twitched slightly.

The light that Carrington saw in the sunspot group was so bright he thought it must have been a hole in his equipment. But this wasn't the case − everything was in order. Realizing that something very unusual was happening, he ran to find someone to share the brilliant sight with, and even though he returned hastily, within just a few minutes the light had moved across the sunspot group and was quickly vanishing − leaving no sign of what had just happened. As Carrington looked on, the appearance of the sunspots was just as it had been before the flash of light. Richard Hodgson, another amateur astronomer, who was observing sunspots from London, also saw the fleeting intense light and these corroborating observations were to prove crucial in getting the astronomical community to take what had happened seriously. Nothing like this had been witnessed before.

Carrington and Hodgson had been the first to witness what we now call a 'solar flare'. What ravaged the Earth later that day was an intense geomagnetic storm, the magnitude of which we thankfully haven't seen since. At the time, because there was no physical understanding to link the geomagnetic storm to the flare on the Sun, Carrington was not prepared to rule out its being a coincidence. We now know that the Sun was to blame and, more than that, it poses a serious threat to our way of life on Earth. Understanding and predicting these events is directly linked with what Carrington was trying to do all along, study the rotation of the Sun.

In a spin

The Sun rotates about an axis like the Earth does and this had
been realized from the early sunspot observations. For observers
in the northern hemisphere it rotates in a right-handed sense.
Close your right hand into a fist, with your thumb pointing up,
and your fingers point in the direction that the Sun rotates and
your thumb represents the axis of rotation and points north.
Sunspots then move from left to right across our view.

Carrington used the movement of sunspots to track the rotation of the Sun's photosphere. We see the legacy of his work
today in our use of the 'Carrington rotation'. This is a twenty-
seven-day rotation rate, seen from the Earth, that Carrington
worked out from sunspots near the Sun's equator. Carrington
rotations are an extremely useful way to track the Sun's rotation
given that there are no fixed features on the photosphere which
would enable lines of longitude to be tracked. The first Carrington rotation was designated as beginning on 9 November
1853 and we have been counting them ever since. As I type we
are in Carrington rotation number 2162.

But it is not as simple as just looking at the sunspots near the
equator. It had been noted, even as early as 1630 by Christoph
Scheiner, that sunspots near the equator move faster than those
up towards the poles. So Carrington was able to use sunspot
motions to quantify speed variations in the latitudes where spots
formed. And by the time of his measurements it had been discovered that there were intriguing regularities and patterns in
the formation of sunspots.

The first pattern spotted in the behaviour of sunspots was a
serendipitous discovery. The Sun was actually being used as a
backdrop to try and find unknown planets that are orbiting
closer to the Sun than the Earth. Anything that passes between
the Sun and us will be silhouetted against the luminous disc and

will be seen scuttling across it. Venus does this twice (eight years apart) roughly every 115 years. Mercury, which is much closer to the Sun and orbiting once every eighty-eight days, crosses much more frequently. This is why Kepler's original interpretation for the spot he saw was that it was Mercury transiting the Sun.

In the 1800s there was a popular idea that there might be a planet orbiting the Sun even closer than Mercury. Like all the planets in the Solar System, Mercury makes an elliptical orbit around the Sun rather than following a perfect circle; its distance varies from 69.8 million to 46.0 million kilometres. The point at which the planet is the furthest from the Sun is known as the 'aphelion' and the closest approach known as the 'perihelion'. Over time the elliptical orbit of a planet drifts around the Sun like a hula-hoop – we say it 'precesses' – and astronomers could see Mercury's perihelion and aphelion drifting around accordingly.

Only they were not moving the way people expected. The perihelion and aphelion points were moving faster than the calculations predicted they should. Factoring in all of the effects from the Sun and other planets predicted a precession in Mercury's orbit far slower than what was actually happening. These calculations had been done meticulously using Newton's equations. It was reasoned that either Newton's equations were wrong, or there must be an unseen planet that had not been included in the calculations.

A precedent had already been set in this area after Herschel had discovered Uranus. Its orbit also was not exactly like that predicted by Newton's theory, so another planet was posited to exist that would have the right gravitational pull on Uranus to explain its motion. A French mathematician, Urbain Le Verrier, was one of the people who calculated exactly what additional planet could be added to the Newtonian model to give the observed orbit of Uranus. Astronomers turned their telescopes to that location and sure enough it wasn't long before the planet we now know as Neptune was found in 1846.

Flushed with his success, Le Verrier turned his maths to the orbit of Mercury. He predicted that there was another planet orbiting even closer to the Sun and he named this hypothetical planet 'Vulcan' after the Greek god of fire. Astronomers went to great lengths to be the first to spot Vulcan.* The difficulty was in working out what was a sunspot and what was a potential planet passing in front of the Sun. Both cases would look like a black dot.

One person who was on the hunt for the new planet was Samuel Heinrich Schwabe, a German pharmacist turned astronomer. He patiently observed the Sun on every day that he had clear skies, year after year from 1826 to 1843. He duly recorded all the sunspots. He knew that a planet would stand out as a spot that repeatedly came back, again and again, whereas sunspots would randomly appear and disappear. But, alas, his decades of work produced no evidence for the hypothetical planet that promised to keep Newton's theory on its scientific pedestal. However, Schwabe's perseverance paid off in a way that he

---

* There is a twenty-first-century analogy to the hunts for Neptune and Vulcan: the search for dark matter. This search was triggered by unexplained observational characteristics of galaxies. These are collections of hundreds of billions of stars that are bound together by gravity, but the stars at the edge are rotating far quicker than our current understanding of physics predicts. Our best models show that the galaxies should fly apart, but this is something that we don't see happening. Once again, either our theories are wrong or there is more matter than we can see, holding the stars in place. This extra matter is too dark to see, and has been given the creative name 'dark matter'. Dark matter is an attractive solution because it is simpler to think that there is more matter to be found than to instead consider rewriting the laws we think govern the Universe.

couldn't have expected when he made a major discovery about sunspots.

As Schwabe looked through his vast record of sunspots a pattern started to emerge. Their number seemed to change over time and did so in a regular way. Schwabe realized that over roughly a ten-year period the number of sunspots rose and fell as if it were following a cycle. His careful and long-term observations also showed that sunspots didn't appear in a random way on the Sun – there seemed to be a repetitive process at work. Schwabe had discovered, and published in 1843, what is now known as the 'sunspot cycle', or 'solar cycle', and it represents the very slow heartbeat of the Sun.

Although the length of the cycle can vary between eight and fifteen years, the average cycle length is found to be almost exactly eleven years. The very early sunspot observations were used to reconstruct the sunspot number back to 1755 and this revealed a cycle that ran between 1755 and 1766. This is known as 'cycle 1' and the following cycles have been numbered consecutively. I started my career in solar physics in cycle 23 and we are currently enjoying cycle 24.

In the decades following the discovery of the sunspot cycle, in Germany the astronomer Gustav Spörer became interested in understanding the distribution of the latitudes at which sunspots formed. At the start of a solar cycle, the first spots form at mid-latitudes around 30–40 degrees above and below the equator. As the cycle progresses, spots start to appear more frequently but the latitude at which they form moves closer and closer to the equator. This is now known as 'Spörer's law'.

As for the planet Vulcan, sadly it was never found. But it turns out that the Sun's spinning had more to do with Mercury's orbit than anyone expected.

Working in shifts

We had better fill in the gaps for how the rest of the Sun is spinning. The regions between 40 degrees either side of the equator were revealed by the movement of sunspots, but outside these bands sunspots are absent. A new method is needed to work out how the featureless plasma is moving in those regions of the photosphere.

In 1842, an Austrian mathematician and physicist, Christian Doppler, developed an area of physics that solved the problem. He realized that the wavelength of a spectral line would be shifted if the material that was emitting or absorbing the light was in motion. If the material was travelling towards us, the line would be shifted to a shorter wavelength and higher frequency, such as towards the blue end of the visible spectrum. If the source was moving away from us, the spectral line would shift into the red, or long-wavelength and low-frequency, end of the spectrum.

This change in frequency of a wave because of motion now has the generic name of 'Doppler shift'. It is an effect we are used to in sound: a siren on a vehicle approaching us sounds higher in pitch than when it recedes from us; racing cars produce the classic high then low engine noise as they race past. Detecting this shift in the wavelength of spectral lines turns telescopes into interplanetary speed cameras.

This Doppler shift could be tested out on the Sun. Firstly, the shift in spectral absorption lines around the equator exactly matched the rotation speed calculated independently by Carrington using the movement of sunspots. Secondly, looking at the left and right sides of the Sun indicated that the plasma was coming towards and going away, respectively, exactly as you would expect from a rotating sphere.

When this technique was used on the plasma up towards the

poles on the Sun it was shown that those regions were moving far slower than at the equator. What was first noticed using sunspot observations continues at higher latitudes. The Sun rotates fastest at the equator and then that speed drops off towards the poles. Imagine the Earth if countries near the equator raced around the globe faster, and had a shorter day, than those at higher latitudes! The question now was, if there is differential rotation on the surface of the Sun, what could possibly be going on inside?

Our ability to see inside the Sun began with observations made at the Mount Wilson Solar Observatory, but some time after the death of George Ellery Hale in 1938. In the post-Hale years, solar physics had become rather repetitive at Mount Wilson. Observations had become routine and there was perhaps little innovation, until a physicist called Robert Leighton arrived. An experimental physicist who did work across all aspects of astronomy and even in early particle physics, he was a friend of and collaborator with the Nobel laureate Richard Feynman. Leighton built and used incredibly accurate Doppler shift solar cameras to reveal that the Sun had another kind of motion. It was constantly vibrating, like the surface of a bell that is being hit repeatedly by a hammer.

The granulations in the photosphere moved around in the chaotic manner you would expect from convection cells, but layered over this was a much more regular movement. The surface was oscillating, with patches rising and falling continuously and regularly. Curiously, everywhere, the oscillations took on average five minutes to vibrate once. It was these oscillations at the visible surface that became the stepping-stone for probing the rotation inside the Sun.

Almost ten years after their serendipitous discovery, it was suggested that the photospheric oscillations were caused by sound waves trapped inside the Sun – the hammer that strikes the Sun is inside – so that the Sun constantly rings like a very low-frequency bell – so low in frequency that it's well below the range of our hearing. And the theory that the Sun was full of sound waves held up to scrutiny – the observations matched the details of the theory – which meant that astronomers now had another way to investigate the Sun: as well as using light they could use sound. And the sound waves are a product of the constant motion of the plasma inside the Sun.

Inside the convection zone, the plasma is always moving and expanding and contracting, and that means that the Sun is acting like a bell which is constantly being struck with tiny blows. Imagine a bell that has sand constantly poured on it rather than being given one large hammer blow. But the Sun has its own natural filter that removes much of this noise and leaves series of sound waves at what are called the 'resonant frequencies'. It is these sound waves that cause the photosphere to vibrate at specific frequencies, and there are millions of them.

Much of what I have learnt about the sounds of the Sun has come from the group that does research in this area at the University of Birmingham. They have been studying the Sun's sounds for many years, and to collect their data they operate a series of telescopes around the world that form a network to continuously monitor the Sun twenty-four hours a day. The network is called BiSON, for Birmingham Solar-Oscillations Network. And using sound to study the Sun is an area of solar physics called 'helioseismology'.

The first person I met in the group was Bill Chaplin and I was pleased when I was able to interview him for the iconic BBC programme *The Sky at Night*. For the interview he took me to the university's Great Hall. As soon as we walked in, I was faced with a vast pipe organ that stands either side of the hall. It has an impressive set of pipes, ranging in size from long fat ones several times as tall as I am, to shorter and skinnier ones. He asked the organist to play a note and a low-pitched, or low-frequency, boom came out. Instinctively I knew that the note was coming from one of the larger pipes.

Bill was making the point that the frequency of the sound that a pipe creates tells us something about the size of the pipe. They too have their own resonant frequencies. The air entering my ear from the sound that the pipe made was probably vibrating at around 100 times a second, which gives it a frequency of 100 hertz. The vibrations of the plasma at the surface of the Sun are much slower, taking place on a timescale of about five minutes (or about 0.003 hertz). But these so-called 'five-minute oscillations' are just part of a much larger family of vibrations and some take up to an hour to make one oscillation.

These frequencies are far below the range that the human ear can detect, which goes from roughly 20 to 20,000 hertz. The sounds of the Sun would need to be speeded up tens of thousands of times if we wanted to audibly enjoy the solar symphony. But it's not only the size of the pipe that matters – the gas inside is important too. The organ pipes that I was standing next to with Bill were full of air at the same temperature and pressure as the air around us in the hall. Inside the Sun the gas is in the form of plasma and its pressure and temperature increase with depth into the Sun. This has a very useful effect on the sound waves: as they travel into the Sun, the increasing plasma temperature speeds up the deeper part of the wave more than the part in the shallower plasma; this bends the wave back up to the photosphere, where we can observe it.

The millions of tones that ring inside the Sun produce a complex pattern of oscillations at the photosphere that are detected by the Doppler shift of the light emitted from the moving plasma. So we do not listen to the Sun in a literal sense: rather, the signatures of the trapped sound waves are coded into the Doppler shifts, and by measuring these shifts we get access to the information carried by the sound.

Measuring these delicate shifts is not an easy task – it requires meticulous observations and the careful removal of the signature of motions caused by convection. But, once extracted, this web of oscillations is used to reconstruct the sound waves that produced them. All this effort is worth it because the trapped sound waves carry information about the interior layers that they have travelled through. This is how they become useful as a probe for the inside of the Sun.

This is not that different from how we know what is going on inside the Earth. Seismic tomography uses the sound waves produced by earthquakes to deduce what is going on inside our own planet. This is how we know that the Earth is more complicated internally than I implied at the start of this chapter – with various fluid shells moving differently.

Helioseismology allowed the internal plasma flows to be probed directly for the very first time when it was realized that sound waves travelling in the direction of the Sun's rotation (with the moving plasma) would have a slightly higher frequency than similar sound waves travelling against the direction of rotation (swimming upstream). Bill Chaplin told me that what they found was not what the models had predicted.

In the outer parts of the core and the radiation zone the solar plasma is rotating almost like a rigid sphere. This shell of plasma extends out to over 550,000 kilometres from the centre and is spinning about once every twenty-five days – more slowly than the photospheric plasma near the equator, but faster than the photospheric plasma towards the poles. So around the equator, the convection zone is sliding over the radiation zone, constantly overtaking it. However, at the poles, the convection zone lags behind the radiation zone spinning below it. (See plate 5.)

You may have noticed that above I used the phrase 'outer parts of the core'. This was very deliberate. In the innermost part of the core, it becomes very hard to use the technique of helioseismology to find out about the rotation rate. Very few of the sound waves make it this far into the Sun and it's hard to pick out the information about the rotation rate of the core and distinguish this from the rotation rate of the higher layers of plasma that the sound wave has travelled through. So for now we can only say that the plasma at the very centre may rotate as a rigid sphere at the same rate as the plasma in the outer part of the core and the radiation zone, but it may not. The movement of the innermost region of the Sun's energy-generating core remains a mystery and an important challenge for helioseismology to solve in the future.

The successes of helioseismology created a new view of the solar interior and showed a plasma sphere that has two sections which rotate in very different ways — a rigidly rotating interior on top of which sits a layer with a differential rotation. The outermost layer is constantly slipping and sliding over the innermost one. This finding was once again totally unexpected. And the region where the rotation transitions from varying across latitudes to being that of a rigid sphere, just below the convection zone, has been given its own name — the 'tachocline'.

This name is inspired by the ocean's thermoclines. In an ocean, the thermocline lies below the top layer of water, which is heated by the Sun and mixed by waves so that it has a fairly constant temperature with depth. Below this lies the main body of cold water. At the interface between the warm upper layer of water and the cold region is the thermocline and the temperature varies greatly across this relatively thin layer. In the tachocline, it is the rotation rate of the plasma that is varying very rapidly across a narrow region.

Helioseismology quickly showed itself to be a very powerful technique in probing the interior plasma flows of the Sun. It came to be used not only to see inside the Sun, but also to see all the way to the other side so that images of the locations of sunspots could be created before that side of the Sun rotated into view. The magnetic fields of sunspots affect the propagation of sound waves. Sound waves that converge on the distant sunspot are modified slightly, causing a change in the frequency of the wave. In this way, sound waves which have bounced from the near side to the far side to the near side again will carry with them information about sunspots that are hidden from our view. I find it amazing that daily maps of the sunspot activity on the far side of the Sun can be created by using sound waves to look right through the Sun.

Proving Einstein right

With the planet Vulcan never having been discovered, a new solution to Mercury's orbit was proposed. In the early twentieth century, Einstein was changing the face of physics with his new ideas, including an overhaul of Newton's theory of gravity. Using his upgrade of Newton's equations predicted the orbit of Mercury to be exactly as it was observed. Einstein's theory was that the perihelion position of Mercury should advance over time because of the strong curvature of space-time close to the Sun.

Except it wasn't the only non-Vulcan solution to the Mercury problem. The Sun was also in the running.

The Sun is formed into a sphere by its own gravity, which is constantly pulling the plasma towards the centre. As gravity pulls, a sphere forms, but it might not be a perfect sphere. The rotation of the Sun brings into play another force that can change its shape: centrifugal force. You feel this force when cornering in a car, and we saw earlier how it gave us the flat disc of the Solar System when it collapsed from a cloud of dust. If the Sun spins fast enough it will bulge around its equator, and if it bulges sufficiently this will also affect the Sun's gravitational field and contribute to Mercury's peculiar perihelion movement. This would mean that the contribution proposed by Einstein's theory was less than predicted and the near-perfect match between theory and observation would be lost. If a bulge in the Sun could be detected, it would throw Einstein's theory into question.

In short: Einstein's theories matched the observed orbit of Mercury, which Newton's didn't. But Einstein knew about Mercury's peculiar orbit in advance. A cynic would say it's easy to make a theory fit data you already have. The upshot of Einstein's work was that it would only work perfectly if we had a bulge-free Sun, and that was something we didn't know in advance. No one had ever measured the Sun's bulge.

However, this is an extremely challenging task. Precise and accurate measurements of the width of the Sun versus the distance from pole to pole need to be made to see whether or not the visible disc of the Sun makes a perfect circle in the sky. This was simply not possible with telescopes. By then it was known the bulge scientists were looking for would only be a few tens of kilometres. This is tiny compared to the 696,000-kilometre radius of the Sun.

To this day, the bulge of the Sun has not been confirmed or denied by telescope observations – and it seems that the exact shape of the Sun varies during the solar cycle. But helioseismology has come up with an answer because it can probe how fast the Sun is spinning on the inside. For a start, the Doppler shift of the photosphere of the Sun had shown that the outer layer was not spinning fast enough for it to bulge out. Any bulge must come from rotation deeper within the Sun. Using sound waves to probe the Sun's interior has revealed that the radiation zone – the rigid ball of plasma which contains the majority of the Sun's mass – is not spinning enough to bulge either. Einstein was right after all, adding weight to a new description of gravity that has been ushered in.

So, in this way, the study of the Sun across the ages, up to and including the modern helioseismology work at the University of Birmingham, is not just for esoteric reasons. It has had direct impacts on other areas of physics. Few modern theories have been put to as much practical use in our modern society as the work done by Einstein. But I started this chapter talking about a practical motivation for understanding the Sun that is a bit more frightening: its ability to disrupt our way of life on Earth.

That also comes down to the movement of plasma within the Sun. Now we know how the parts of the Sun rotate, and we understand that magnetic fields form sunspots, we can combine the two to explain the dramatic event that happened in September 1859, when the skies lit up with aurorae and some mysterious effect paralysed the electric telegraph. What we'll see next is that the rotating plasma inside the Sun is powering a magnetic dynamo, which amplifies the Sun's magnetic field and stores energy in it. This energy can be released in the Sun's atmosphere to drive the biggest explosions in the Solar System. And they have grave implications for humankind's future. Our deadline to understand the Sun is becoming more urgent.

7. The Dynamic Sun

If the Sun had no magnetic field, it would be as uninteresting as most astronomers think it is.

Attributed to Robert Leighton, *c.*1965

This quotation attributed to Robert Leighton, father of the five-minute oscillations, often comes to my mind when I am talking about the Sun. I love it because I can see that the Sun probably does appear unremarkable when it is pitted against other astronomical objects like super-massive black holes, distant galaxies or the mystery of dark matter. With cosmic wonders like these, how can a seemingly bright circle with a few occasional dots compete?

Well, the Sun comes to life when you realize it has a magnetic field. It is the presence of this field that turns an otherwise bland ball of hydrogen into a star which has spots and flares and which changes from day to day and month to month. And as we saw with the Carrington event, the Sun is an electromagnetic body that can release enough energy for it to directly interact with our own magnetic field here at home, on Earth. To understand these solar attacks, we need to answer three questions:

1. Why does the Sun have a magnetic field?
2. How strong is the magnetic field?
3. Why is its magnetic field so complicated and variable?

Incredibly, the answers to all these questions come down to the thin convection zone on the outside of the Sun, not the massive interior below, where the fusion is happening and which actually powers the Sun.

Origin of the magnetic field

Despite only being able to see less then 5 per cent of the Universe directly, observations reveal that magnetic fields are everywhere. Stars have them, galaxies have them and even clusters of galaxies that are bound together by gravity have magnetic fields threading between them. One scenario for these cosmic magnetic fields is that they were created during the first 400,000 years after the Big Bang, by the freely moving electrically charged particles − electrons and protons − which the Universe was made of at that time. We see that magnetic fields pervade the Universe and we think that they may have been there for a very, very long time. For this reason, it is possible that these primordial magnetic fields were still there when the Sun formed.

Magnetic fields don't last for ever but cosmic magnetic fields can have very long lives. If the electric current that is creating them dissipates and disappears, perhaps because collisions between the charged particles interrupt its flow, so too does the magnetic field. Size matters too. When you turn off your kettle, the electric current stops and the magnetic field goes to zero. For a kettle, this is almost instantly. For plasma structures on an astronomical scale, it can take much, much longer. And this could explain the origin of the Sun's magnetic field.

It is conceivable that the cloud of gas and dust out of which the Sun formed had its own weak magnetic field threaded through it. If some of the gas that makes up the nebula was ionized − electrons having been stripped away from the atoms − as the nebula collapsed, the ionized gases would have dragged the magnetic field with them. Not only did matter coalesce and clump together to form the Sun, but magnetic field could have been piled up as well.

From the size and electrical conductivity of the Sun it's expected that any magnetic field that was there when the Sun formed from the solar nebula might take billions of years to decay. So there could be some remnant still trapped inside the Sun today. But that would be far too weak to match what we see, and definitely not enough to explain the strong magnetic fields inside sunspots. And the remnant of the magnetic field could be expected to form a simple field with magnetic poles close to the rotation poles – as we have on the Earth. But even though the Sun and the Earth share some similarities when it comes to their magnetic fields, there are many differences to what we see at their respective surfaces. We need to explain a magnetic field on the Sun that is very complex as well as being very strong, forming sunspots that vary in number and position at the photosphere.

It's worth remembering that the presence of a magnetic field in the Sun isn't surprising since the Sun is made of plasma. After all, moving electrically charged particles are the origin of all magnetic fields, and the spinning Sun has an abundant supply of these. The plasma of the solar sphere makes the Sun a great conductor of electricity on a colossal scale. A plasma is one of the best electrical conductors there is. So when electric currents start flowing inside the Sun, they don't meet any resistance and it's very hard to stop them. As long as the electric currents keep flowing, the magnetic field will remain.

But there is an added complication. A magnetic field embedded in a plasma is nothing like the magnetic fields we are used to at home. When you attach a magnet to your fridge, or turn on the kettle, you are using a permanent field locked in a bar magnet or inducing a new one with a flowing current. A plasma falls somewhere in between.

A plasma is not a solid object which has moving electrons within it (either moving within atoms or electrons flowing down a wire) but rather a mobile mist of its own charged particles. And it is vast. In your home the magnetic field created by your fridge magnet or the current in your kettle's wire emanates out into the air in the room. In the Sun, the magnetic field spreads through the plasma, which is moving all the time in a multitude of directions. A branch of physics has been developed to describe the interaction of magnetic field and a plasma: magnetohydrodynamics.

This branch of physics allows us to probe how a magnetic field changes over time when it threads through a mobile plasma. As we might expect, this area of physics has a long history and is the outcome of people's curiosity about electric and magnetic fields. To understand how magnetic fields change over time requires a combination of discoveries that have been applied to electrically conducting fluids like the Sun. They are named after the people who worked on them. Combining Ampère's law to describe the magnetic field around the electric current that is flowing, Ohm's law to describe the electric fields associated with a mobile plasma and a magnetic field, and Faraday's law which describes the relationship between magnetic field and electric fields, gives rise to a curious consequence.

As well as the mobile particles in a plasma moving and creating a magnetic field, the magnetic field becomes 'frozen in' to the plasma. This is the result of how the forces applied to a charged particle in a magnetic field strangely always act orthogonally to the field lines (called the 'Lorentz force', after the Dutch mega-physicist Hendrik Lorentz). The field can have as much control of the plasma as the plasma has of the field.

This exclusively sideways force means the particles are never pushed further away, or drawn closer to the magnetic field, but rather are made to spiral around the field lines, forever trapped like a planet around a star. And either one can take the other with it. The plasma cannot move without shifting the magnetic field as well and the magnetic field will never let the plasma escape. This is why the intense magnetic fields of sunspots trap plasma in place and allow it to cool.

The task now is to use this knowledge of magnetohydrodynamics to link the original simple magnetic field the Sun was born with, and that probably lives on in the radiation zone, with the magnetic mess we see coming out of the plasma-filled convection zone. The leftover primordial magnetic field by itself would simply be too weak to form the surface magnetic features of sunspots. But if the Sun formed with a weak magnetic field already present, it could have acted like a seed that has grown and become the magnetic field that we see today. Somehow the plasma takes the magnetic field the Sun was born with and mutates it into something much more interesting.

Origin of the magnetic strength

It's always a good approach when trying to find an answer to a tricky question to start with a simple scenario and gradually build in the complexity. So let's ignore the complex magnetic fields of the sunspots for now, and just think of the Sun as having a simple global magnetic field, much like we have here on Earth. The Earth has magnetic north and south poles that are close to the geographic poles about which the Earth is spinning. And the Sun has a magnetic north and a magnetic south pole that are both very close to the axis about which the Sun rotates. The only difference is that the polar magnetic field of the Sun (at the photosphere) is about twenty times as strong as the Earth's polar magnetic field at the surface. So far so good.

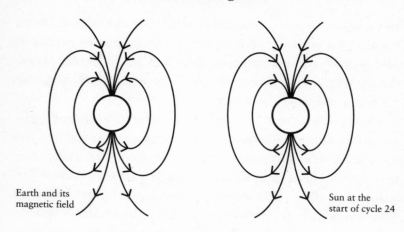

Earth and its
magnetic field

Sun at the
start of cycle 24

7.1 A schematic showing the configuration of the Earth's magnetic field
and the Sun's magnetic field (at the start of cycle 24). Not to scale!

In both cases we can imagine the field lines coming out of the
north pole, curling out through space and then re-entering at
the south pole. The magnetic fields of the Sun and the Earth are
not that dissimilar to giant bar magnets. All we have to do is
close the field lines; as we know, they always form closed loops.
For a bar magnet they run straight back up the middle. Within
the Earth the magnetic field goes up through the liquid metallic
(mainly iron and nickel) outer core and over to the opposite
pole. But because the Sun spins at different speeds at different
places inside, things get much more complicated.

Our starting magnetic field has field lines that run parallel to
lines of longitude and this configuration is known as a 'poloidal
field'. If the Sun had no interior plasma flows and the Sun rotated
as a rigid body, the field lines inside the Sun would always run
parallel to lines of longitude, like they do in our very simplified
description of the Earth. And this is the case for any magnetic
field in the core or the radiation zone, where everything rotates

in unison. But in the convection zone, where helioseismology has shown that the plasma flows vary with latitude, known as 'differential rotation', there is a very important consequence for the magnetic field. In the convection zone the plasma exerts enough pressure on the magnetic field to drag it with the flow. The internal magnetic field can be *distorted* by the plasma flows.

It's this ability of magnetic fields to be distorted that I find so fascinating. It's as if they are elastic and flexible – in the Sun, magnetic fields can evolve in ways I never imagined possible. Before I studied the Sun, magnetic fields to me were fairly straightforward. Pass a current through a wire, and you create a magnetic field in the region around it. I could wave my hand around the wire and move the air but this would have no effect on the magnetic field. Turn off the current and the magnetic field disappears. Shortly after starting in solar physics I had to give up all my intuitions.

Now I have an office full of pipe cleaners and elastic bands that I use with my colleagues to visualize the shapes in the magnetic fields that can come about when the field is threaded through an electrically conducting gas, and how the shapes can change when this electrically conducting gas is moving. I became interested in understanding how this shifting plasma environment might be responsible for evolving a seed magnetic field in ways that lead to sunspots and the sunspot cycle.

Inside the Sun the plasma flows act as fingers that grasp the magnetic field lines. The rotating plasma at the base of the convection zone (in the tachocline) pulls more on the magnetic field at the equator because that's where the plasma is rotating faster. Day by day, the plasma fingers stretch the magnetic field and the field lines slowly get drawn out like an archer pulling on the string of a bow. Eventually, after about eight months, the field lines have all been dragged along so much that they wrap all the way around the Sun!

At this point, the field lines catch up on themselves at the point where they started. And they then overtake the point where they began. Instead of one magnetic field line in this location, there are now two. Rotation after rotation, the magnetic field lines are wrapped around and around the Sun and very tightly wound bundles of magnetic field form in both hemispheres, which connect over the equator. This magnetic field configuration is known as 'toroidal' as the shape of the magnetic field looks like two toruses, or doughnuts — one in the northern hemisphere and one in the south.

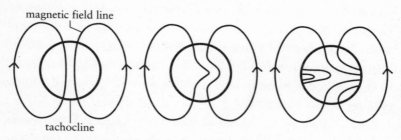

7.2 The rotation of the plasma inside the Sun draws out the magnetic field in the tachocline region and starts to wrap it around. This turns the magnetic field from being 'poloidal' to 'toroidal' in its configuration.

In this simple scenario the rotation of the plasma has two important effects on the magnetic field. First, the orientation of the field lines inside the Sun has changed. They started off being parallel to lines of longitude and ended up being closely aligned to lines of latitude. And, second, each time a field line wraps more than once around the Sun it has the effect of amplifying the magnetic field strength. Presumably the Sun has been doing this ever since the convection zone developed these sheared flows. Magnetic field lines do not weaken as they are lengthened by the moving plasma. The small seed magnetic field has been stretched, folded and grown into one that can explain the strength of the Sun's electromagnetic influence we see today.

Origin of the magnetic complexity

We now have an amplified magnetic field at the base of the convection zone, but that does not explain how the magnetic field responsible for sunspots appears at the photosphere. It seems that the magnetic field which causes sunspots to be dotted over the Sun's surface has its birth down at the tachocline: the region between the radiation zone and the convection zone where the differential rotation can keep amplifying the magnetic field for the longest time. So how, then, does the magnetic field formed at the base of the convection zone form sunspots at the photosphere?

The short answer is simple: the magnetic field floats to the surface. In reality, the process is much more complicated. As the toroidal magnetic field grows, where the magnetic field wraps around itself the most, it weaves together to form 'flux ropes', complete with plasma now trapped inside these ropes. A flux rope is actually somewhere between a rope and a hose. It is rope-like in that there is magnetic field distributed right through it; it's not hollow. But it is hose-like in that plasma is trapped in it, only allowed to flow along the flux rope and not leak out of the sides.

The magnetic field in a flux rope really does start to act like it is made of elastic bands. A flux rope has an aspect of springiness to it. This means that if the plasma outside the magnetic flux rope pushes on it, the flux rope can deform inwards, responding to the plasma pressure outside. But if the flux rope is squashed inwards, it means that the flux rope becomes more densely packed with magnetic field lines, and this has an important effect because it increases the magnetic pressure in the rope. The magnetic field presses on the plasma *within* the rope and it starts to move along it and thins out. With the aid of the magnetic field, the plasma inside can be less dense and yet the flux rope can still have the same pressure as the ambient plasma around it.

This means that within the flux ropes there can be patches of plasma that become less dense than the plasma outside the rope. We are not sure why certain sections develop this low density before others, but once the plasma starts to thin out in one part of the rope it can become buoyant. And like anything that is less dense than the fluid around it, it will start to float to the surface. But instead of dragging the whole flux rope up, one part of it will bow out and start to stretch up towards the surface of the Sun. As the buoyant plasma rises, it carries the magnetic field with it and forces the magnetic field to start to resemble the shape of the Greek letter omega, Ω, as part of it stretches away, making a bid for the photosphere. Eventually the top of the omega loop penetrates the photosphere and, as it crosses, a pair of sunspots form.

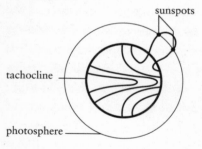

7.3 Sunspots form when a loop of magnetic field rises from the tachocline and bursts through the photosphere.

The ongoing weaving and release of flux ropes at the tachocline explains why we see pairs of sunspots in the Sun's photosphere. Each pair represents the two locations where the loop of magnetic field crosses the photosphere. Instead of a relatively contained magnetic field like that of the Earth, the Sun's magnetic field is constantly building up and spilling out of the sides. It is these highly strung flux ropes that caused the catastrophic event witnessed by Carrington. But there are still a few other sunspot mysteries to clear up before we come back to that.

Magnetic mysteries

As soon as it was possible to measure the magnetic field within sunspots, something interesting was noticed about the pairs. In each pair, one sunspot seemed to be 'leading the way' around the Sun. On average, a line drawn between leaders and followers shows a very slight tilt of about 4 degrees on average away from the east–west line. This tilt doesn't depend on which hemisphere the spots are in. This tilt of the sunspots is known as 'Joy's law', after Alfred Harrison Joy, who worked at the Mount Wilson Observatory.

Hale wrote in 1919: 'The following spot of the pair tends to appear farther from the equator than the preceding spot, and the higher the latitude, the greater is the inclination of the axis to the equator.' The preceding spot refers to the spot which is at the front in relation to the rotation of the Sun. It's the one on the right-hand side of the pair. And, more than that, the leading sunspot of the pair always has the same magnetic polarity for spots in the same hemisphere: either always a positive magnetic polarity or a negative. But if the leading spot of the pair in the northern hemisphere was positive, then the leading spot in the southern hemisphere would be negative. Across 1735 sunspot pair observations, Hale only found forty-one that did not have the same leading magnetic field direction.

Even more interesting was that each time a new cycle started, the magnetic polarity of the leading and following polarities

switched. Two sunspot cycles needed to play out before the Sun
returned to its original magnetic configuration. The Sun may
have an eleven-year sunspot cycle, but its magnetic cycle is
twenty-two years in length. In solar cycle 23 the leading sunspots in the northern hemisphere were of a positive magnetic
polarity (negative magnetic field led in the south); in cycle 24
the leading spots have a negative magnetic field (positive magnetic field leads in the south). The Mount Wilson data were
throwing up lots of observations that needed an explanation.

We'll start by explaining the tilt in sunspot pairs named after
Joy. On the way floating up from the tachocline there are several
competing effects that can influence the ascending flux tube.
The effect responsible for misaligning the eventual sunspot pairs
is actually one we are familiar with here on Earth: the Coriolis
force. This is the force imparted on anything moving within a
larger spinning system. Because both the Earth and the Sun are
constantly rotating, everything that moves within them experiences a Coriolis force.

The Coriolis force works in different directions, depending
on your hemisphere, and is often thought to determine the way
in which your water spins down the plughole, but in reality the
water can spin either way because the Coriolis force is too weak
compared to any movement the water may already have from
when the bath or sink was filled. But, even so, scientific research
has been done on whether the Coriolis effect is seen when
emptying a sink. A paper was published in the prestigious journal *Nature* in 1962 that set out to test this hypothesis. A special
circular tank was designed and built by the engineer Ascher
Shapiro at MIT that was 1.8 metres across and 15 centimetres
deep. The tank was filled with water and left to settle for twenty-
four hours. The plug was then pulled and in the first fifteen
minutes of draining no rotation was seen in the way the water
left the tank. But in the last few minutes a counter-clockwise

rotation was seen, the direction expected for the northern hemisphere. A flurry of similar experiments followed, including a
test in the southern hemisphere, which showed the water rotating clockwise as it drained.

Were these results concrete proof? The answer isn't clear. The
Coriolis force only really comes into play when things are moving over vast distances. On the Earth this explains why hurricanes
and cyclones spin in opposite directions in different hemispheres.
Within the Sun it explains why flux ropes, or more accurately
the plasma trapped inside, rotate in slightly different directions in
different hemispheres as they float to the surface. But even with
structures as massive as flux ropes, they only spin by around
4 degrees across their entire journey. And in forty-one of the
cases Hale observed, the buffeting from the convection currents
or other flows swamped the Coriolis force completely.

More interesting, though, is trying to explain why the activity on the Sun is not constant. This is perhaps the most exciting
part – the flows within the convection zone do not simply keep
amplifying the magnetic field so that it grows and grows and
grows. This isn't what we see at the photosphere. Instead there
is an ebb and flow – the magnetic field pulses in strength as it
changes from sunspot minimum to sunspot maximum. Somehow it is being regenerated and reconfigured from one cycle to
the next. And these changes aren't restricted to the sunspots –
they reach throughout the Sun.

The magnetic field in the polar regions reverses in magnetic
polarity and it does this at the same point in each and every solar
cycle – around the maximum of the solar cycle, when sunspots
are most abundant and when the spots are forming at roughly 15
degrees from the equator in the northern and southern hemispheres. Could the link between the peak in the sunspot number
and the polar reversal somehow be connected? Given that the
internal changes that come from the different rates of plasma

rotation aren't able to affect changes at the poles, it's not straight-forward to see how sunspots at such low latitudes could have consequences so far away. To probe this question needs a more detailed look at the photosphere and exactly what is happening to the magnetic fields that emerge there and, once again, at how the plasma is flowing.

All spots are transient features only blighting the photosphere for days or weeks. Only the largest, rare, spots will survive for several months. But with 400 years of observations, we have very detailed images of what happens to sunspots in the lead-up to their disappearance from the photosphere, when they start to fade away, and they all show the same evolution. The initially coherent dark annulus splits apart and hot plasma from the sur-rounding photosphere breaches the spot. Bridges of light start to form that connect from one side of the spot to the other. The spot keeps on splitting and smaller and smaller fragments form.

So, over time, the once coherent tube of magnetic field that formed the sunspots is slowly broken up and subdivided into smaller and smaller magnetic tubes which separate from each other and start to spread over a larger and larger area. The sun-spot disappears from our view as the magnetic field breaks into pieces and is no longer able to trap enough plasma to cause it to cool. But, even though there are no sunspots, the magnetic field is still there.

The ongoing presence of the magnetic field at the photo-sphere is a crucial point. The sunspot does not decay in the sense that the magnetic field ceases to exist – we've already seen that the Sun is such a good conductor of electricity that the currents which are maintaining the magnetic field still flow. What's hap-pening is a *redistribution* of the magnetic flux over the photosphere, and the important mechanism that moves the magnetic field around is the convection which we see as supergranulation. These flows sweep the small bundles of weak magnetic field around the photosphere. And since there is no constant pattern to the convection flows – the cells constantly appear and disappear all over the photosphere in a random way – the magnetic fields are randomly dispersed over a larger and larger area.

Is this gradual dispersal enough to produce changes at the poles? If the very small magnetic fragments of the dispersed sunspots are tracked, another subtle but important flow is revealed that means the fragments might be able to reach this far.

Following the movement of the magnetic fragments and the plasma reveals that there is a flow from the equator towards the poles and this acts as a giant conveyor belt for the remnant sunspot magnetic field. Such a poleward conveyor belt was first considered in 1925 by Eddington when he was thinking about how a star might redistribute its plasma to maintain shells of constant temperature within the star. He realized that circulation currents would flow in meridian planes (along lines of longitude) – called 'meridional flows' – that make giant conveyor belts in the outer layers.

But it was not until the development of helioseismology that we could properly probe these flows. This technique was used in the rise phase of cycle 23 to show that the photospheric poleward flow extended into the Sun for about 26 million metres, 4 per cent of the solar radius. And they extend to more than 75 degrees in latitude in each hemisphere. In this zone the plasma was flowing towards the poles at around 10–20 metres per second – a difficult flow to detect against the random surface motions that move the plasma at speeds of 1000 metres per second and the rotation of the Sun which carries the plasma at 2000 metres per second. But how do these flows reverse the magnetic field at the poles?

This is where the slight tilt of each sunspot pair comes into play. The twist of the Coriolis force means that the trailing sunspot is closer to the pole already, and in the majority of cases the trailing sunspot is of the opposite magnetic polarity to the nearby pole. So the plasma 'conveyor belt' preferentially drags up the magnetic field to a pole that is of the opposite magnetic polarity to that spot. And, if enough magnetic field could be carried up, it could bring about the magnetic field reversal seen around times of solar maximum.

The solar dynamo

So now we have a fairly complete model of how the Sun can take its original simple magnetic field and turn it into something amazing and dynamic. This whole system of the plasma flows and movement of the magnetic field is called the 'solar dynamo'. To recap:

The simple magnetic field at the bottom of the convection zone is stretched out by the moving plasma and gradually wound around the Sun. This amplifies it and causes flux ropes to form. If the plasma trapped inside these flux ropes becomes less dense in one area, there can be a run-away effect where that part of the rope becomes buoyant and starts to float towards the surface, forming an omega shape. It finally breaks through the surface and forms sunspots where it intersects the photosphere. But, on the way up, Coriolis forces twist the tube, meaning that the trailing part of the tube is slightly closer to the Sun's nearby pole. Fantastically, the trailing sunspot is of the opposite magnetic polarity to the pole, so when its magnetic field is distributed over the photosphere, and the solar-pole conveyor belt takes its magnetic field up to the pole, it is taking a magnetic field which has the opposite polarity to what was initially there. This accumulates and eventually swamps the previous polarity, causing the magnetic poles to systematically swap. Finally, in turn, this means the next generation of flux ropes forming at the base of the convection zone and later emerging will be the other way around (magnetically speaking) and the cycle continues, for as long as these flows are maintained.

Phew.

It is perhaps lucky for those studying the Sun that its magnetic field is variable on timescales that can be appreciated within a human lifetime. The dynamo-driven solar cycle lasts on average 11 years and 36 days. This gives us enough time to collect the data over many cycles, enabling us to understand what processes are at play. I have been studying the Sun long enough to start measuring the passing of time in units of solar cycles. I started my career in solar physics in cycle 23. I watched as the north pole flipped its magnetic polarity in late 2000 and as the south pole reversed its polarity in 2001. There is no reason why they should flip simultaneously if this process is indeed driven by magnetic flux dispersing from sunspots in the respective hemispheres. Sunspots do not form as mirror images in the northern and southern hemispheres. And the meridional flows in the north and south do not have to be exactly the same either.

I eagerly waited to see the Sun pass into cycle 24. This cycle finally began in January 2008, when the first spots of the cycle were seen in the northern hemisphere. Cycle 24 was a surprisingly slow cycle to start, showing us that despite the advances we have made in understanding the dynamo, the Sun still has the ability to act in a way that we don't expect. Not all cycles are the same size and it's hard to predict how strong the upcoming cycles will be.

And we still have many questions to answer. For example, why does the sunspot cycle last eleven years on average? Why not one thousand, or one million years? Or why not a few months? And in fact the eleven-year cycle isn't the only pattern that can be pulled out of the sunspot data. The modern understanding is that several cycles are overlaid on top of each other – it's just that the eleven-year cycle is the most obvious one.

Some modern ideas are that a memory may exist between solar cycles. The output of one cycle may lead into the next cycle or the cycles after that. The meridional flow may vary significantly from one solar cycle to the next. It could be that changes in the flow speeds have an effect on subsequent solar cycles. For example, if the meridional flow speed increases, more magnetic field is transported across the Sun and the next cycle might be longer.

Conversely, if the conveyor belt slows down, other processes, such as differential rotation or diffusion of the magnetic field, have more time to come into play. Then it is a case of which of these competing effects dominates to influence the development of the field and the strength of the cycle. Even though nuclear fusion defines our sun as a star in that it generates light, it is the plasma motions that control the dynamo and define the magnetic character of the Sun.

The key thing here is that the solar cycle, driven by the dynamo, continually brings magnetic flux into the solar atmosphere. It prompts the question: what happens to all this magnetic field? Does it endlessly accumulate over time? And what are the consequences for the Sun's atmosphere? To answer these questions we need to look at the atmosphere of the Sun into which all the magnetic field is emerging. The first way we can do this from the ground is during times of a total solar eclipse.

8. Eclipses and Rainbows

On 22 July 2009, I stood on the deck of a ship sailing eastwards across the Pacific Ocean. I was on my way to see a spectacle that was being heralded as the astronomical event of the century. As the Sun and the Moon dance around the sky, there are occasional chance alignments when the disc of the Moon completely blocks the light coming from the Sun, producing a total solar eclipse. The eclipse of July 2009 would be the longest such alignment this century, creating a full 6 minutes and 39.4 seconds of darkness for those people in the right place on Earth.

Solar eclipses always attract attention and people will travel around the world to see them. There were 1500 eclipse chasers on my ship, all waiting for those moments of darkness. But the Moon does not actually block all of the light from the Sun and I was there to see the small amount of light that the Moon cannot hide. This light reveals the Sun's atmosphere – the plasma layers above the photosphere. And there was a very personal reason why I wanted to witness this event.

After having studied the Sun's atmosphere for a decade this was going to be the first time that I would actually look at it directly. The Sun's atmosphere is normally hidden from our view because its faint light is lost in the dazzling glare of the photosphere. But this isn't the case during a total eclipse. The Moon is big enough to block the photosphere so that we cannot see light coming from the Sun's visible surface, but it's not large enough to prevent us from seeing the atmosphere. For the previous ten years I had relied on specialist telescopes above the Earth's atmosphere to view the Sun's atmosphere in wavelengths that our eyes cannot see. But now I was about to witness an event that reveals the normally hidden solar atmosphere at a time when it's perfectly safe to see without any specialist equipment. I would see the atmosphere with my own eyes.

The best location to view the totality of an eclipse varies from event to event and can be almost anywhere on Earth. So I wasn't complaining when I found out that this one was best viewed from a cruise ship on the sunny Pacific Ocean. On the morning of eclipse day, we sailed past the volcanic island of Iwo Jima, which was the location of a bloody battle during the Second World War and has now reverted to its pre-war name of 'Iwo To'. It looked so peaceful surrounded by the calm ocean that morning. But, as noon approached, the calm was broken as the ship's deck became thronged with people. The eclipse was imminent.

As I stood looking at the Moon sliding into position, I couldn't help but marvel at the fact that total eclipses occur at all. They only happen because of an astronomical coincidence. Not only does the Moon pass in front of the Sun, but it is also exactly the right size to cover it up. Actually, the Moon passing in front of the Sun is not that surprising if you remember how the Solar System was formed: the Earth, Moon, Sun and other planets formed from the same flattened nebula disc, so they all orbit in the same plane.

Well, almost. If the Moon orbited the Earth in *exactly* in the same plane as that in which the Earth orbits the Sun, we would see a solar eclipse once every lunar orbit. Unfortunately for solar eclipse fans, the Moon's orbit is tilted by about 5 degrees to our orbit around the Sun. So the Moon tends to pass above or below the Sun from our point of view. Only occasionally does it happen to go directly between the Earth and the Sun.

The truly lucky coincidence, though, is the relative size of the Sun and the Moon. The Sun is 400 times larger than the Moon but it is also 400 times further away from us. So from our perspective both are the same size in the sky. This means that when the Moon moves in front of the Sun it is just the right size to cover it up. If the Moon were smaller/further away it would not block all of the Sun from our point of view (hence annular eclipses occur at the Moon's apogee); if it were bigger/closer it would block the Sun's atmosphere as well as the photosphere.

We actually live at the perfect time in the life of the Solar System for this alignment to work. The Moon is moving away from us at a rate of about 4 centimetres every year, about the same speed that your nails grow – imperceptible to us but, given enough time, the consequence of this will be significant. It means that if there are any humans on the Earth in a billion years' time the Moon will be too small in the sky to produce a total solar eclipse. For our distant descendants, eclipses will be seen as a black disc that covers most of the Sun but not all. Or maybe humans will have moved to a new planet orbiting another star with better eclipses by then!

Even when everything is perfectly aligned, the Moon only casts a small shadow on the Earth. For any total eclipse, the Moon blocks the view of the Sun for about just 1 per cent of the Earth's surface. And the shadow that is cast on the Earth is moving! Because the Moon is moving and the Earth is rotating, the shadow sweeps across the planet. So to see an eclipse you need to be in the right place at just the right time. You have to travel to see a solar eclipse; it will rarely come to you.

The central part of the shadow, the umbra, sweeps along what's known as the 'path of totality' and there will be a location along this path where the total eclipse has the longest duration. I was waiting to see the eclipse at this point on the Earth where the total eclipse would last for 6 minutes and 39.4 seconds. The position and bearing of our ship, which was chasing the Moon's shadow, were to give us an extra three seconds of totality. Totally worth it.

Eclipses through the centuries

I find it remarkable that we can talk about eclipse positions, timings and durations down to a fraction of a second. It shows just how detailed our knowledge of the motion of the Moon around us is, and our motion around the Sun. We can predict exactly when an eclipse will occur and how long it will last and how fast the shadow will be moving. I had an eclipse map based on these predictions and it showed me exactly where and when the Moon would cast its shadow. These maps are vital for allowing people to get to the right place at exactly the right time. Remarkably, eclipse maps are not a new thing.

The first time an eclipse map was used to tell the public about the location of the Moon's shadow was in 1715. The map was created by Edmond Halley, the British mathematician and astronomer who is famous for his work on the orbits of comets. He even had one named after him. Halley wanted to know how fast the shadow would sweep over the Earth, but to work this out he needed observations from various points along the path of totality. Since he couldn't be at more than one place at once, he was going to need some help.

Thankfully the 1715 eclipse was going to be visible across England and Wales, meaning there would be plenty of people who could help. It was the first eclipse to be observed in England for 500 years and Halley realized it would generate excitement that he could capitalize on to get some science done. He devised what was probably the first citizen science project and requested that the 'curious' of the country who were along the path of totality observe 'what they could' and make a note of the time and duration of totality from their location.

Halley had to ask the public for help because there were only two universities in England at that time, so there was no large team of professional researchers spread across the country to call upon. And it was just as well that he did call on the public because the astronomy professors at the two universities had no luck in seeing the eclipse. Clouds obscured the view at one site and the Reverend Cotes at Cambridge 'had the misfortune to be oppressed by too much company'! – a common problem for astronomers during eclipses. Enough observations came in, though, and Halley was able to calculate that the shadow swept over the Earth at a staggering 2800 kilometres per hour.

Eclipses are incredibly versatile and have contributed to many areas of science. Just over 200 years after Halley's experiment, the total eclipse of May 1919 was used to investigate Einstein's new ideas about how gravity was the result of massive objects distorting the shape of space-time. Once again, it was Eddington who observed this eclipse so that he could test Einstein's ideas, and when he gave a lecture at the University of Cambridge about what had happened he inadvertently inspired Cecilia Payne to become an astronomer.

In short, Einstein's theory of general relativity made predictions about how massive objects, like the Sun, should bend the path of light passing by. This means that the stars we see near the Sun should actually appear in slightly the wrong place as the light coming from them is bent by the Sun's mass. But this was impossible to check normally, because the Sun is so bright we can't see stars near it in the sky. Only during a total solar eclipse would it be possible to check the locations of the stars right next to the Sun and see if they were ever so slightly displaced, as predicted by Einstein. This was exactly what Eddington planned to do.

Eddington did not get to go on a cruise in the Pacific though; the 1919 eclipse was best viewed on the African island of Principe. But he did get to see an even longer eclipse than the so-called 'eclipse of the century' I got to see. The 1919 eclipse totality lasted a full 6 minutes 51 seconds, beating mine by around 9 seconds. And thankfully it was a very long eclipse as the weather on the day was terrible. Only briefly during the almost seven minutes did the skies clear enough for Eddington to take a group photo of the Sun and nearby stars. Before 1919, Einstein's ideas were unverified and seemed too strange to be true. But that one photo was all it took to change this: the stars were indeed in slightly different places. It looked like Einstein's theories had been confirmed.

Totality

At 26 minutes and 40 seconds past eleven on the morning of 22 July the Moon slid perfectly into place and completely covered the Sun. Up until then, everyone had been looking at the Sun through 'eclipse glasses'. These are glasses made from the solar filters you can put on telescopes (the ones which are darker than sunglasses or even welding masks). The Sun is so bright that even if it is partially covered by the Moon it can still cause damage to your eyes if you look directly at it. Only when the Moon is completely blocking the Sun is it safe to do this.

So the moment had arrived: I could finally look without my eclipse glasses. Around me the temperature had fallen noticeably without the direct heat from the Sun. The colours of the darkened sky had changed and taken on an unfamiliar purple hue and part of the horizon had the appearance of sunrise. But I didn't care about any of this. I was looking up at the Sun. And there it was, a wispy halo that surrounded the occulted Sun: the Sun's atmosphere.

I was struck by how extended it is – 99 per cent of the Earth's atmosphere is held, by gravity, below 55 kilometres from the Earth's surface. That distance is less than 1 per cent of the Earth's radius. For the Sun, I could see its atmosphere extending out to a distance of around 700,000 kilometres, which is roughly equal to the Sun's radius. The physical extent of the Sun's atmosphere is much greater than you might expect.

The light I was looking at coming from the Sun's atmosphere is a million times fainter than the photosphere, giving it a pearly appearance. And in the temporarily dark sky the planets Mercury and Venus were suddenly revealed. And, just like Eddington, I could see a small number of stars too. As I stood there marvelling at the eclipse, I was following a long line of astronomers who had done the same. For centuries, the only time that humans could see the Sun's atmosphere was during a total solar eclipse.

But the fragile wispy halo I could see was only one part of the Sun's atmosphere, a part called the 'corona', so named because it sits like a crown around the Sun. Staring up at it, I could see exactly why it was given that name: it did look like a majestic crown. The other layer of the Sun's atmosphere that I wanted to see wasn't as obvious as the corona, as the Moon was still hiding some of it.

Halley managed a glimpse of this other layer when he viewed the eclipse of 1715. Just as totality was ending, Halley, who had been using his six-feet-long telescope, saw that as the photosphere began to reappear he didn't see the bright white light that he was expecting. Instead, he had a fleeting glimpse of a long but narrow crescent of deep red light that arced round the edge of the Moon. In seconds it was gone and the photospheric light lit up the daytime sky once again. What on the Sun had he just seen?

It turns out that what Halley saw is a layer of the Sun's atmosphere which is the stepping stone between the photosphere and the corona. Its distinct rosiness is in stark contrast to the white light of the photosphere and the corona and has earned it the name of the 'chromosphere', meaning sphere of colour. It spans several thousand kilometres above the photosphere but can only be seen with the naked eye just as totality begins and ends during a total eclipse. The Moon is big enough to cover the chromosphere, and so you only see it when the Moon is not quite in position (but close enough to being in position for it still to block the light from the photosphere).

Eclipse observations showed that the chromosphere is an irregular and jagged layer of the atmosphere. It has a shape resembling the spines on a hedgehog, vertical tubes that have become known as 'spicules' that extend upwards for 9000 kilometres, a distance that is not far off the diameter of the Earth but only a tiny fraction of the size of the Sun. Jets of hot plasma shoot up the spicules and they live for no more than fifteen minutes or so before fading away, only to be replaced by new ones. And then there are clouds of chromospheric plasma that are lofted up to great heights, rising 150,000 kilometres into the corona above and visible when the Sun is fully eclipsed. The Victorian eclipse observers described them as 'red mountains'. Today we call them 'prominences'.

Thankfully the brief flash of the chromosphere at the beginning or end of an eclipse is all that is needed to capture an image of the chromospheric spectrum. As always, we learn about the plasma in different parts of the Sun by studying the spectra that they emit. At first glance there are some similarities between the spectra of the chromosphere and of the photosphere, in that a very faint rainbow of continuous colours is seen in both. But there are two major differences.

Firstly, the spectrum of colours is much dimmer in the light coming from the chromosphere. The whole rainbow is there, but it has been muted. Secondly, there are no dark Fraunhofer lines in the spectrum of the chromosphere. In place of the dark lines there are, bizarrely, extra-bright lines! When the light coming from the chromosphere is split into its spectrum the cosmic barcode is reversed: what was dark becomes bright and what was bright is diminished. So why does the chromosphere produce a spectrum that looks so different?

We know the dark Fraunhofer lines appear because light from the photosphere has to pass through the rest of the photospheric plasma as it travels out and some of its photons are stolen along the way. The photosphere gets cooler with height and some of the atoms and ions in the relatively cool gas are able to absorb photospheric photons, but only those photons with just the right wavelength (frequency) to make an electron jump up to a higher energy level. This gives us the sharp absorption lines in the otherwise continuous rainbow.

Importantly, though, when these atoms and ions absorb a photon, that energy is not destroyed or otherwise disposed of. The electron can drop back down from this higher-energy orbital and the same frequency photon can be re-emitted. But that released photon can go in any direction, not necessarily travelling on towards us like the light it was originally stolen from was. So while most frequencies can stream out of the photosphere, those few frequencies that can be absorbed and re-emitted by atoms and ions in the chromosphere have a much harder time, constantly being bounced away from us, and perhaps even being sent back into the photosphere, where they can be absorbed by a negative hydrogen ion so that the whole process has to start again.

But in the chromosphere at the start and end of a total solar eclipse the exact frequencies that struggle to get out of the photosphere suddenly become the frequencies we see the most of! But this is a kind of optical illusion, a trick of contrast.

Normal Fraunhofer lines are not actually completely dark. The photons at those frequencies are absorbed and re-emitted in random directions which still allow them to eventually escape the Sun. It's just in comparison with the other frequencies from the photosphere, which are so much brighter, that these lines are made to look dark. When we view the chromosphere during an eclipse, we are looking high enough in the Sun's atmosphere for, from our point of view, the photosphere not to be behind it. So when the photosphere is completely blocked by the Moon, we only see these selected frequencies being scattered sideways out of the chromosphere and towards us. So they seem relatively bright.

Now we can see what these lines in the spectra tell us about the chromosphere. The most prominent line in the chromospheric spectrum is the hydrogen alpha line. This line is in the red part of the spectrum and it's what gives the chromosphere its trademark red colour. This was the wavelength in which Hale chose to image the Sun in 1908 and it allowed him to see the plasma in the chromosphere just above the sunspots outside of the time of a total solar eclipse and not pick up the whole spectrum of colours of the light coming from the photosphere. This is why the spectroheliograph that he used was so important: it filtered out all of the light from the photosphere, leaving only the faint chromosphere. It was the same effect seen during an eclipse but, instead of only being at the very edge, it worked right across the surface of the Sun.

If you do look at the Sun using just the light of hydrogen alpha it is incredibly beautiful, the swirling prominences the Victorians had seen around the edge of the Sun actually extending right across it. Prominences appear as dark features known as filaments and can be tracked across the Sun as it rotates. They are actually clouds of relatively dense and cool chromospheric plasma suspended in the corona. In hindsight, the name 'red mountains', which the Victorians used for prominences, was surprisingly appropriate: they contain around the same mass as a mountain on Earth − 100 billion kilograms. They are vast structures that are somehow held aloft by an invisible force. Gravity isn't able to pull these plasma mountains inwards.

1. A glass plate from the Harvard College Observatory archive which has tiny stellar spectra recorded on it. An eyeglass is needed to magnify and interpret the spectra.

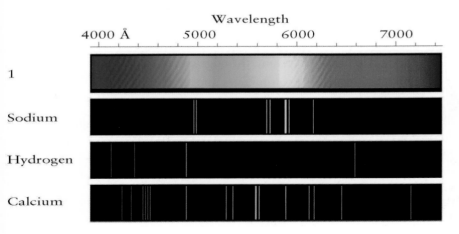

2. The rainbow colours of the so-called continuous spectrum (*top row*) contrasted with the emission line spectra of the thin energized gases sodium, hydrogen and calcium, which show individual lines at certain wavelengths (© *21 May 2015 OpenStax College*).

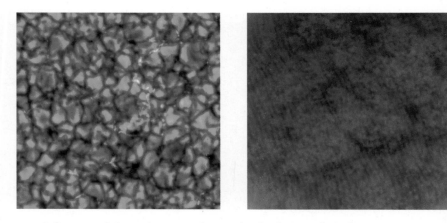

3, 4. *Left*: image of photospheric granulation taken by the Hinode satellite (*JAXA/NASA/UKSA*). *Right*: miso convection. Miso soup, that is.

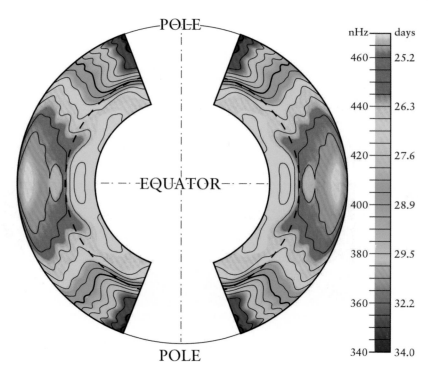

5. A diagram showing how fast the plasma inside the Sun is spinning. There is an obvious change in rotation rate between the bottom of the convective zone and the top of the radiative zone (*GONG and SOHO/MDI consortia*).

6. Image of the Hale–Bopp comet, which was visible in 1997. The dust tail stretches out to the right, while the bright blue ion tail is pointing almost directly away from the Sun (*ESO/E. Slawik*).

7. Extreme ultraviolet light image of the Sun's corona taken by NASA's Solar Dynamics Observatory satellite. Active regions appear bright whereas coronal holes, where plasma escapes the Sun as the fast solar wind, appear dark. Superimposed on the image is a model of the Sun's atmospheric magnetic field. White lines indicate magnetic arches whereas brown lines show the 'open' field lines that extend out into the Solar System. These are the magnetic super-highways (*NASA/SDO, AIA science team and LMSAL*).

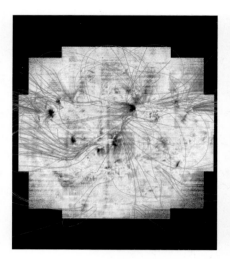

8, 9. Images of the solar corona in extreme ultraviolet light. The image on the *left* is a so-called Doppler image where the plasma flows are shown (towards us in blue and away from us in red). Superimposed in green and orange are the magnetic field lines from a computer model. On the *right* is an image of the Sun showing in red, orange and green the regions where the solar wind escapes (*Dave Brooks/JAXA/Hinode EIS team/Nature*).

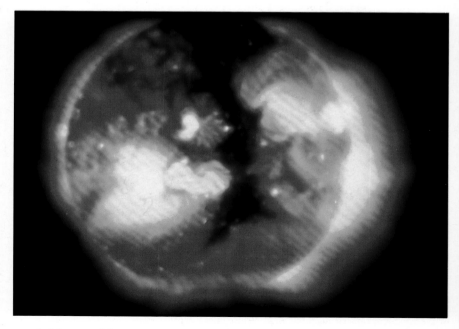

10. Skylab view of the corona in X-rays in 1973. These images revealed that the corona is highly structured (*NASA*).

11. Skylab debris at the Esperance Museum in Western Australia. In front of me is an oxygen tank and above my head a copy of the cheque from a US radio station that paid for the littering fine issued to NASA!

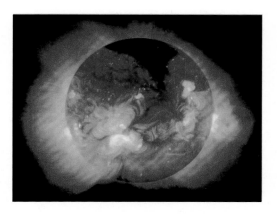

12. A glorious image of the corona glowing in X-rays taken by the Soft X-ray Telescope on the Japanese Yohkoh satellite in May 1992 (*JAXA/ National Astronomical Observatory of Japan/University of Tokyo/LMSAL*).

Apr 21 2002 03:56:22

13. A solar flare 'slinky' seen in extreme ultraviolet radiation. This image was taken by NASA's TRACE satellite in April 2002 (*NASA/Goddard Space Flight Center Scientific Visualization Studio*).

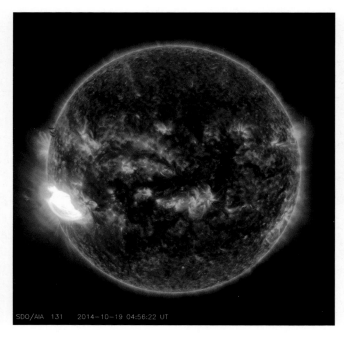

SDO/AIA 131 2014-10-19 04:56:22 UT

14. Extreme ultraviolet image of the Sun showing a solar flare on the left side. Image taken by the AIA telescope on NASA's Solar Dynamics Observatory (*NASA/SDO and AIA science team*).

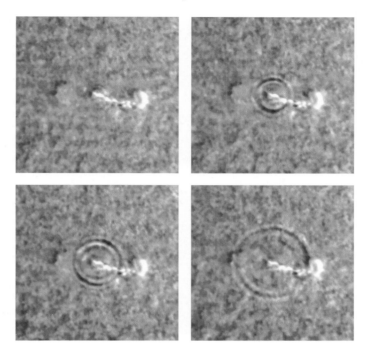

15. The concentric rings of the first sunquake to be detected. The rings are produced as sound waves ripple up to the photosphere. The data were taken by the MDI instrument on the SOHO spacecraft (*ESA/NASA/Alexander Kosovichev/Valentina Zharkova*).

16. The 'light bulb' coronal mass ejections as seen by the SOHO spacecraft. The size of the Sun is shown by the small white circle (*ESA/NASA and the LASCO consortium*).

2000/02/27 07:42

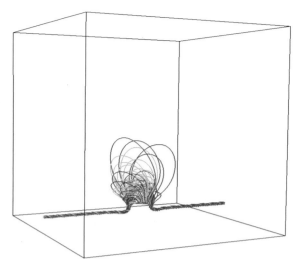

17. This image shows a simulated magnetic flux rope that has partially emerged through the photosphere. The magnetic field lines are shown in blue. This simulation has been created by Vasilis Archontis of the University of St Andrews – a centre of excellence for modelling the solar magnetic field.

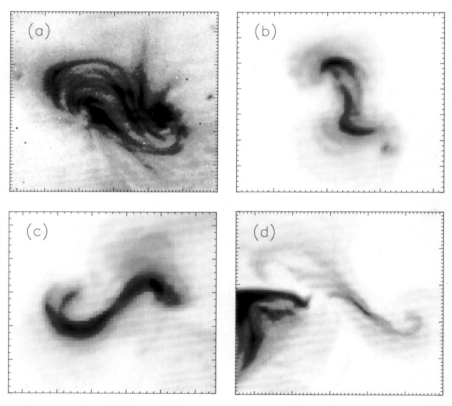

18. A variety of so-called 'sigmoids' seen in X-ray images of the corona (*JAXA and the SXT consortium/Lucie Green*).

The new elements

The chromospheric spectrum shows more than just the inverted Fraunhofer lines: there are extra lines – other wavelengths of light are emitted that are not seen in a normal solar spectrum. In the nineteenth century a spectral line was spotted in the yellow part of the visible spectrum that doesn't appear as a Fraunhofer line and didn't match anything that had been produced by a gas studied in the laboratory. Scientists couldn't work out what element was responsible for this new line.

The same thing happened in the corona. When the technique of spectroscopy was used to study the faint light coming from the corona it wasn't long before another bright emission line, which hadn't been seen before in the laboratory, was observed. By the end of the nineteenth century, three other unidentified coronal emission lines had also been discovered: a green line at a wavelength of 503.29 nanometres, a yellow line at 569.45 nanometres and a red line at 637.45 nanometres.

As all of these lines had never been seen using elements on Earth, there was only one conclusion: there must be new elements on the Sun! Thirty new elements had already been discovered during the nineteenth century so it must have seemed entirely plausible that this was the explanation. The mysterious new element in the chromosphere was called 'helium' after the Greek personification of the Sun, Helios, and the new element in the corona was called 'coronium'. You cannot help but notice that one of those elements is more familiar-sounding than the other . . .

In our modern world of helium party balloons, it is hard to imagine a world where helium was completely unknown. But that was the case in the 1800s. The discovery of the chromospheric spectral line assigned to helium came independently to the French astronomer Pierre Janssen and the British scientist and astronomer Norman Lockyer in 1868. Both were very adept at astronomy and spectroscopy and both had developed techniques to study the chromosphere's light.

Janssen was also an early eclipse chaser. After the eclipse of 1868 he went on to travel to see eclipses in 1870, 1875, 1883 and 1905. They took him all over the world. To get to Algeria for the eclipse of 1870 he had to find a way out of his hometown of Paris, which was surrounded by Germans at the height of the Franco-Prussian war. He made his escape in a hot-air balloon (yet another situation when helium would have helped).

Lockyer in his later studies stayed closer to home, his pioneering work in the area of spectroscopy and astronomy being reflected in his appointment as the first ever university professor in 'Astronomical Physics' at what is now Imperial College, London. It was Lockyer who suggested the name 'helium', and for the next thirty years the Sun was the only place where it had ever been found.

It wasn't until 1895 that helium was finally discovered on Earth, when an experiment with a radioactive uranium mineral was carried out at my university, UCL, showing once again that the material that makes up the Sun is also the material that constitutes the Earth. The experiment was carried out by William Ramsay, who was head of chemistry at the time and one of the most famous scientists of his day. He discovered five new elements, which have become collectively known as the 'noble' gases, was knighted in 1902 and was awarded the Nobel Prize for Chemistry in 1904.

We still have at UCL today the original sample that Ramsay used to make his discovery, its having been rediscovered after many decades in storage. From this discovery and the work of Cecilia Payne-Gaposchkin we know today that helium makes up around 25 per cent of the mass of the Sun, and in fact it is the second most abundant element in the Universe, having been formed along with hydrogen after the Big Bang. So why was the second most abundant element so hard for us to find that it was discovered on the Sun before we found it right on our doorstep?

Following the solar nebula model for the formation of the Solar System, the Earth should have had its fair share of helium. But that original stock has since been lost. The reason for this, and the reason why helium is used in party balloons, lies in the small mass of the helium atom: the gas particles are simply too light and move too fast for the Earth's gravity to be able to hold on to them. Like many an accidentally released party balloon, the early helium on the Earth simply drifted up and away.

What we have now is 'new' helium formed within rocks during their slow radioactive decay. All the helium you see in balloons has actually been mined.

The Sun kept its original helium but it is hard to *see* and the reason for this is also a consequence of the very small size of the atom. In order for helium to reveal its presence on the Sun it must either absorb a photon and contribute to an absorption spectral line or emit a photon and contribute to an emission line. But the smallness of the helium atom, where the two electrons are held tightly to the nucleus, means that absorption lines are hard to produce. There is actually very little radiation emitted by the Sun with enough energy to do this. The result is that there are no helium absorption lines in the photospheric spectrum for our eyes to see.

For a helium atom to emit a photon the atom must have an electron that has already been promoted to a higher energy level, perhaps through a collision with a freely moving electron that has been liberated from another particle. When the electron falls back to its original energy level it releases the energy gained by emitting a photon of a very particular wavelength and this contributes to an emission line. The problem here is that the plasma needs to be at around 20,000 Kelvin for the freely moving electrons to be moving fast and have enough energy to be able to do this. In the photosphere the plasma simply isn't hot enough to excite the helium atoms and cause them to radiate light. Yet helium emission lines are seen being emitted by the chromospheric plasma, and this reveals something important – the plasma in the chromosphere must be hotter than that of the photosphere.

The observation of helium in the chromosphere not only meant the discovery of a totally new element but it also showed that going up in altitude from the photosphere, moving further away from the energy source in the centre of the Sun, the temperature of the plasma starts to increase rather than continuing to decrease. This is completely counterintuitive. Up until this layer the temperature of the plasma has been dropping off with distance from the core because energy is being lost through radiation at the photosphere. Now that situation is reversed. There must be another energy source that is heating the chromosphere. (We'll come back to this later.)

Coronium has a very different story and sadly never found its way into party balloons. Mainly because it does not exist. One of the other great inventions of the nineteenth century was the periodic table, in which all elements could be categorized. It was so powerful that gaps in the table even make it possible to predict the existence of previously undiscovered elements. The hunt was on for coronium to fill one of the gaps.

By the late 1800s the periodic table was filling up, though, and there was less and less room for coronium to fit in. It was becoming apparent that a previously undiscovered element might not be the cause of the new spectral lines in the corona. Seventy-four years after coronium was first discovered it was found that it does not exist – it was 'anti-discovered'. It turns out that coronium was actually iron and calcium in disguise.

The green coronium line is produced by the emission of

photons from an iron ion that has lost thirteen of its electrons (Fe XIV). The yellow line is formed by the emission of photons from a calcium ion that has lost fourteen of its electrons (Ca XV) and the red line from iron with nine lost electrons (Fe X). These particles have all lost a significant number of their electrons.

The lines were difficult to identify because the laboratory conditions in which incandescent gases were being studied were so vastly different to the plasma of the corona. The density of the corona is only 10 million billionths of that of water, or 10,000 millionths of the air around you, and such a thin gas cannot be reproduced easily in the laboratory. This means that the ions in the lab can behave differently to those same ions in the Sun. And conversely it shows why the Sun makes such a fantastic laboratory in its own right – we can study an environment that we can never re-create here on Earth.

In the laboratory, changes to the energy that an ion has are controlled by collisions with electrons – the gas is always dense enough for collisions to be occurring. In the low-density corona the frequency of collisions is low. If a collision between an electron and an ion does occur it will be a long time until the next collision takes place. During this long wait, the ion can spontaneously release the energy it previously gained by emitting a photon. The exact wavelength of the photon depends on the ion and how much energy it has, but the possibilities include the green, yellow and red emission lines. The very low likelihood that these emission lines would be seen on Earth earned the name 'forbidden' lines. Actually, they are not forbidden, they are just extremely rare.

This explained coronium but produced a new mystery. For so many electrons to have been knocked out of iron and calcium atoms, the plasma temperature must be very high. If scientists had been shocked to discover that plasma in the chromosphere had a temperature in the 10,000s Kelvin, they were about to be blown away by the corona. For iron and calcium to produce the coronium lines, the plasma in the corona must be over a million Kelvin. That's several hundred times hotter than the photosphere.

Crowning glory

Looking up at the corona from the cruise ship, I could not see any of these spectral details. To me the corona looked its normal pearly white. But even that reveals something about the nature of the corona. The thin plasma of the corona shouldn't be able to produce a continuous spectrum – that's only something that a much denser gas can do.

The hint to what is going on is that the continuous coronal spectrum has a very similar appearance to the photosphere's spectrum. This tells us that the photons that we see during an eclipse coming from the corona were created in the photosphere. When I looked at the eclipsed Sun in 2009, it was actually *photospheric* light that I was seeing. The light originates in the cooler, denser plasma lower in the atmosphere. But as the photons rush outwards from the photosphere they flood into the million-degree coronal plasma where many electrons are flying around that have been stripped away from the parent atoms. In the same way that photons are scattered in the radiation zone by electrons, photons also get scattered in the corona. The ones that are scattered in our direction are the ones that we see.

Then, finally, I looked right out, at heights of about two or three times the radius of the Sun, where a very faint ring of light could still be seen. In this region the photons are still being scattered in our direction, but not by free electrons in the coronal plasma. This light is scattered by tiny dust particles that were left over from the formation of the Solar System from the solar nebula. High above the corona sits a faint souvenir from the birth of the Solar System itself.

Then, as quickly as it started, the eclipse was over – 6 minutes and 42 seconds can really fly by. I had no complaints about my trip to the Pacific, but not all eclipses have been in such great holiday destinations. The early discoveries and the curious observations of the corona meant it wasn't long before a telescope had been invented that mimicked nature and could create an artificial solar eclipse. No longer did astronomers need to travel across the globe for a fleeting glimpse of the corona; they could study at leisure from home.

A French astronomer, Bernard Lyot, invented this telescope, which is known as a 'coronagraph', and in 1930 successfully made the first observations from an observatory in the Pyrenees. In this telescope the image of the Sun was focused onto a circular disc that played the role of the Moon in a real eclipse. The disc was just larger than the image of the photosphere and prevented the photospheric light from going any further into the telescope. The faint light from the corona was not blocked and a further lens was used to focus this light into an image. Much of my job involves looking at images taken with a coronagraph, which is how I am used to seeing the corona – and why it was so great to finally see it for myself.

Now we are left with the problem of why the chromosphere and corona are so hot. It seems the Sun's atmosphere gets hotter the further it gets from the surface, not cooler, as we expect. And this raises the very natural question: does anything come after the corona? Just how far does the Sun's atmosphere extend?

9. Bon Voyage

In 1977 NASA launched two identical spacecraft, both on a two-year mission to study the planets Jupiter and Saturn, with the option of one spacecraft then going on to Uranus and Neptune. The 1970s were an exciting time, when the plans for Solar System exploration became much more ambitious, moving beyond exploration of the Moon, Venus and Mars to the outermost planets. And it was a well-timed trip because the planets Jupiter, Saturn, Uranus and Neptune were about to move into a special arrangement that meant a spacecraft could visit all of them in turn, hopping on from planet to planet. This alignment only happens once every 175 years and NASA had plans to make the most of this rare opportunity. So the Voyager 1 and Voyager 2 spacecraft were launched and, despite being built to survive for five years, they are still journeying through space and sending information back to us today.

The Voyager spacecraft were sent off on their mission from Cape Canaveral, Florida, being launched just two weeks apart. Their expeditions had been designed so that as each spacecraft reached its first planet it would take observations and then be flung on to the next in a gravitational slingshot. The slingshot is a clever way to use gravity to our advantage. As Voyager 1 and 2 approached their target planets, the gravitational pull that they felt from the planet increased and the spacecraft speeded up. The path of the spacecraft bent as it passed close by and for Voyager 1 and 2 this propelled them further out into the Solar System with no need to use their own fuel. It meant that the Voyagers could be sent much further out into the Solar System than the rocket that launched them could do alone.

The Voyagers were designed to journey to the far reaches of the Solar System; their journey was one of epic proportions. Jupiter is just over five times more distant from the Sun than we are – 778 million kilometres – and that means it receives less than 4 per cent of the sunlight that we do. To power the spacecraft in the darkness they carry space batteries that use a radioactive source: plutonium-238. All materials that are radioactive naturally emit particles of energy as the nuclei, which are unstable, reconfigure and rearrange themselves. For plutonium-238 an alpha particle is emitted which is made of two neutrons and two protons – it has the same composition as the nucleus of a helium atom. This pared-down particle of plutonium changes into the element uranium, but it's not the element left behind that is important. The constant emission of the alpha particles generates heat as the moving particles are absorbed into the surrounding material. This heat, also called 'thermal energy', is then turned into electricity.

When the spacecraft were launched these space batteries provided them with the same power as roughly four 100-watt light bulbs. It may not sound like much, but all space missions are designed to run on what we would consider a very frugal energy diet. They have to, in fact, as it would be simply too expensive to equip them with anything more. And this small amount of power is all there is to run everything – onboard computers, heaters, instruments and the system that allows the spacecraft to communicate back to Earth. But the radioactive power also meant that each spacecraft had the potential to survive into middle age despite the plans for a five-year working life.

The visits to Jupiter and Saturn were a huge success and spurred the NASA scientists on. The mission was extended so that Voyager 2 would slingshot on to become the first spacecraft to Uranus and Neptune. But after these visits, with no further planets out there – well, except for Pluto at that time – the primary mission ended after an incredible twelve years. It had been a phenomenal success, during which time the Voyagers had discovered that Jupiter's moon Io is the most volcanically active object in the entire Solar System, made studies of forty-eight different moons around the outer planets, and found that Titan, Saturn's largest moon, has a thick nitrogen atmosphere and looks like a planet in its own right.

Even though the primary mission was over, the two Voyager spacecraft were very much alive and functioning well. The plutonium power source had been declining though, a natural outcome of the radioactive nature of the material. The more time passes, the more plutonium atoms have reconfigured themselves and the fewer there are left to decay by emitting an alpha particle. Radioactive elements are characterized by what is called their 'half-life', which describes how long it takes for the radioactivity to fall to half its original level. Plutonium-238 has a half-life of 87.7 years.

But even though the power had been dropping both Voyagers still kept going. And with some clever adjustments that included turning off some of the instruments, the spacecraft continued to do their science. The work of Voyager 1 and 2 was far from over and a unique opportunity lay ahead. Attention turned away from the planets and in 1989 the Voyagers were given a new mission: the Voyager Interstellar Mission. Their space batteries were expected to remain useful until around 2020, which might give them enough time to leave the Solar System and become the first human-made objects to enter interstellar space. Even though they had left all the planets behind, they hadn't yet left the Solar System. NASA had a very clear idea of what the *edge* is and it actually begins with the Sun.

Tails

Up until now we have been looking at the Sun as a sphere of plasma, which can be broadly split into different shells. Then, during the time of a total solar eclipse, we get a glimpse that the plasma stretches up from the photosphere to form an atmosphere. The impression that we get from the total-eclipse observations is that the atmosphere gets thinner and thinner with height until it eventually fades and runs out of material.

But over the centuries there have been some icy visitors to the inner Solar System which hinted that the tenuous corona we saw in the previous chapter doesn't end where we see it end with our eyes during an eclipse. They gave the first clue that the atmosphere of the Sun extends beyond what our eyes perceive. These icy visitors are the comets.

Comets are frozen lumpy bodies of ice and dust several kilometres across that mostly reside in a region called the 'Oort Cloud', named after the Dutch astronomer who proposed that a vast shell of comets surrounds the Solar System at one fifth of the distance to the nearest star. That's over 10,000 times further out than Jupiter. At that distance, there is no way that we can see any comets directly, not even with the most powerful telescope. But, still, the Oort Cloud is thought to exist. At that vast distance, comets are only loosely gravitationally bound to the Sun. So they can be easily knocked out of their icy ghetto by the gravitational influence of a star passing nearby. This interaction can change the orbit of a comet so that it comes in closer to the Sun. Again, this may seem far-fetched, but during the lifetime of the Solar System, as we fly through the Galaxy, such incredible events can happen. This gravitational disruption is thought to explain why we have comets that repeatedly come and visit us in the inner Solar System, like Halley's comet (named after Edmond Halley, whom we met before), which comes back to us every seventy-six years.

At their distant outpost comets can take millions of years to orbit the Sun, but when one gets dislodged from the Oort Cloud and comes into the inner Solar System, its journey close to the Sun causes it to heat up. Basking in the increasing amount of sunlight falling on it, ice that has been frozen for millennia turns into gas and particles of dust trapped amongst the ice grains are released. As the comet reaches roughly the orbit of Jupiter, a diffuse cloud of gas and dust called a 'coma' starts to form around the giant iceberg. And then something spectacular happens. A beautiful tail forms behind the comet that stretches out for millions of kilometres. And it is this tail that shows something intriguing.

Comet tails all have something in common: their tails always point away from the Sun – within a few degrees – no matter whether the comet is approaching the Sun or whether it has passed around the Sun and is heading back out into the depths of space. So on its way out the comet's tail is actually ahead of it! This tells us that the motion of the comet itself is not the major factor in setting the direction of the tail. The material of the tail isn't simply being laid down behind the comet as if it were leaving a trail behind it. Something is always pushing the tail out from the Sun.

One possibility is sunlight itself. The pressure of light falling on the dust particles in the coma around the comet can provide a slight force that pushes the particles out and results in a tail. There is no mystery there. But comets have another tail, which is slightly different in colour, and this one cannot be blown back by sunlight – it is a tail of ionized gas. (See plate 6.) Gas particles in the comet become ionized by the ultraviolet light coming from the Sun: electrons are knocked off their atoms, which creates positively charged ions. These are then blown back into what is really a plasma tail. But sunlight doesn't have the same effect on electrically charged particles as it does on dust particles. Something else must be washing over the comet that is not sunlight but which is still originating at the Sun.

The only observational clues to work with first of all came from the detailed viewings of the comet tails. They showed that there is a very small difference in angle between a comet's ion tail and the direction directly away from the Sun. This was a key piece of information because this difference could be explained if the comet was cutting through an outflow of some sort, perhaps a wind from the Sun. If the comet were stationary it would form a tail that pointed directly away from the Sun. But if it were moving across something flowing out from the Sun, its trail would go off at a slight angle. Which is exactly what was observed. The Sun must be producing some kind of invisible wind.

Living in the Sun's atmosphere

Work on where the Sun's atmosphere might actually end really accelerated in the 1950s. In this decade, a British physicist, Sydney Chapman, used the fact that the Sun's atmosphere is hotter than its surface to calculate that this would cause the corona to be 'puffed up'. The high-temperature plasma would exert a pressure and so would expand outwards – just like bread rising in the oven because the air trapped in the dough expands as it warms.

This explained the size of the corona seen during a total solar eclipse. But there was more to Chapman's findings. Plasma conducts heat exceptionally well because of its freely moving particles and this means that it can conduct heat, and remain hot, out to very large distances. Chapman realized that its high temperature must be maintained far beyond the corona that is seen during a total solar eclipse. The Sun's atmosphere was bigger than anyone had ever expected.

In fact, Chapman's theory had the Sun's atmosphere extending out to a distance far beyond the Earth. Chapman realized that the Earth orbits *inside* the atmosphere of the Sun. An astonishing thought that shows we are more intimately connected to the Sun than we ever imagined. Thankfully the Earth's magnetic field protects us from the Sun's massive atmosphere as we fly through it; something we'll visit again later.

Chapman's theory took us one step closer to understanding how the Sun can reach far out into the Solar System but it didn't explain what was actually causing the ion tails of comets. A large tenuous atmosphere was not enough to cause ionized tails to blow back off comets. There must be more to the Sun's atmosphere than Chapman's model was able to capture. To explain why comets have ion tails, we had to answer the question: is the Sun's atmosphere static or dynamic? Does it simply extend into space, or is it actively expanding?

In 1958 another revelation came when a controversial theory was developed that was able to unite the ideas about the formation and orientation of comets' tails and the model of a very extended solar corona. And this new model provided an important clue about the Sun's atmosphere. It showed that it isn't possible to have a corona that is static and merely conducting heat out to very large distances. Instead, since the hot plasma is exerting a pressure it means it would be constantly expanding the plasma outwards, against the Sun's gravitational pull.

Close to the Sun the static model appeared to be correct because the Sun's gravitational pull is stronger there. But further out the Sun's gravity weakens and it gets to a point where the Sun cannot hold on to the hot plasma of its atmosphere and it actively expands out into the Solar System, forming a solar wind – just as comets' ion tails indicated. The constant expansion of the corona provides a flow of plasma that would blow as a wind of charged particles into the Solar System. This solar wind would blow over the comets and produce their ion tails.

A few words should be said about the man behind this model: the American astronomer Eugene Parker. His theory really was a game changer and it wasn't easy for the research community to accept this new idea. Parker followed the normal route for disseminating new ideas and wrote his theory up for publication in one of the community journals, the *Astrophysical Journal* (the journal co-founded by Hale). The paper was sent to two referees, who act as a quality control and provide independent advice to the journal that the paper is worthy of publication. But the referees thought the paper had no scientific value and rejected it. Parker's theory was simply too much of a radical departure from the accepted view at that time.

Luckily, the editor of the journal at that point was Subrahmanyan Chandrasekhar, who worked down the corridor from Parker at the University of Chicago. He was an outstanding astrophysicist himself and in his work he showed that stars which have more than one and a half times the mass of the Sun collapse and become unimaginably dense when they die. These objects were later given the name 'black holes'. Chandrasekhar walked down the hall, poked his head into Parker's office and asked if he was really sure about his work. Parker was. Luckily for solar physics, Chandrasekhar ignored the referees.* Written in late 1957 and submitted to the journal at the start of 1958, the paper was finally published in November 1958.

Even though that exchange took place almost twenty years before the Voyagers left the launch pad, it is centrally important to their Interstellar Mission. Parker's model described a wind that could extend the atmosphere of the Sun to 100 times further out than the Earth. The wind would only stop when its own pressure (which would get smaller as it went out from the Sun) became the same as the pressure of the interstellar gas beyond. The solar wind

---

* The referees had actually not found any flaw in Parker's calculations or mistakes in his logic, so there really was no reason to reject the work, but it was still unusual to ignore their decision.

would create a vast plasma bubble within which the Solar System is immersed.

According to this model, the edge of the Sun's atmosphere would be just over twice as far as the distance to the (now) dwarf planet Pluto. This is the boundary that the Voyagers were looking for. Once they moved beyond the bubble, they would be in interstellar space. But to understand exactly where the edge might lie, and how the edge might be detected, needs a good description and understanding of exactly what the solar wind is like. And this needs measurements to be taken from much closer to the Sun.

It didn't take long to test Parker's theory as his work coincided with the start of the space age. Within just a few years spacecraft ventured outside the Earth's magnetic field, meaning that if the solar wind really did exist, they would be able to detect it outside the Earth's own protective magnetic bubble. And that's exactly what happened. The measurements revealed the existence of a solar wind that is gusty and that sometimes blows fast and sometimes slow – relatively speaking, that is. The fast solar wind streams outwards at an incredible speed of around 800 kilometres per second and makes the 150-million-kilometre journey from the Sun to the Earth in just over two days. The slow wind still moves at a cracking pace and has a speed of around 400 kilometres per second, and takes around four days to reach the Earth. Even at its slowest, the solar wind is faster than any wind on Earth. In fact, no wind in the Solar System moves faster.

Parker predicted that as the solar wind blows out from the Sun it creates a bubble we call the 'heliosphere'. For the Voyagers to find the edge of the heliosphere they need to sense and measure the conditions in the surrounding space so that they can identify when the particles around them stop being those of the solar wind and start being those of the material between the stars. Inside the heliosphere is the domain of the solar wind and outside is interstellar space. But there is one more way to characterize where the solar wind ends and where interstellar space starts.

The Sun's magnetic field needs to be considered too. Parker's description of the expanding corona didn't include the effects of the magnetic field. For a complete description and a thorough understanding of the solar wind it must be included. And when it is included, it solves a problem that we met in chapter 3: that the Sun, which formed at the centre of a collapsing and spinning nebula, is revolving 400 times more slowly than we expect. We expect the Sun to be rotating rapidly as a consequence of the same physics that speeds up a spinning ice-skater as they bring in their arms. On the Sun, the solar wind plasma flows out, guided by the magnetic fields, which are spinning with it. This effectively increases the size of the Sun. The outward flow of the solar wind is equivalent to the ice-skater opening out their arms, and it slows the Sun's rotation down. But while the magnetic field might have solved this 'angular momentum' problem, it presents another one for our understanding of how the solar wind escapes in the first place.

Magnetic highways

As we already know, the whole corona is threaded by magnetic fields that emanate from the Sun's interior and rise through the photosphere. And the magnetic field in the corona creates a complex web of structures – some large and some small. Since the magnetic field and coronal plasma are frozen together – particles can move along the lines of magnetic force, but not across them – the escape of the solar wind is much more complex than initially thought. And almost sixty years after Parker predicted the existence of the solar wind, there is still an awful lot that we want to find out.

The fast solar wind is easier to explain. We are getting slightly ahead of ourselves, and I have already said that the solar wind causes some of the Sun's magnetic field to reach out into the Solar System to Earth and beyond. Amazingly, there are magnetic field line-links between the Sun and the Earth! We call them 'open' field lines (in that they do not immediately come back down into the Sun to 'close' the loop) and they act as magnetic super-highways along which the plasma can flow. (See plate 7.)

We know these are the cause of the fast solar wind because we can track the plasma as it moves. The same spacecraft that measure the speed of the solar wind as it flows over them can also measure what particles the wind is made of. And the make-up of elements changes between the fast and slow streams, giving each bit of the solar wind a signature composition. When we use spectroscopy to look at what elements are in the plasma at the base of these open field lines in the corona, it matches the plasma which later washes over our spacecraft as the fast solar wind.

The problem comes with the magnetic fields above sunspots on the Sun's surface (called active regions). We saw before that these are loops of magnetic field emerging out of the photosphere, bending back over and then going back down into the Sun. Any plasma in these magnetic fields is doomed to remain trapped – unable to escape the magnetic field because it is channelled back down to the Sun.

But some must get out. The reason I confidently state that the plasma escapes, despite the apparent paradox, is because we have the measurements to prove it. It becomes the slow solar wind. Measurements of the slow solar wind made by spacecraft sitting in the flow exactly match the elemental signature of plasma that was last seen trapped in the magnetic fields of active regions. By studying the spectrum emitted by this plasma we can work out what it is made of. This is the plasma's great escape.

Some light was shed on this paradox in 2012, when one of my

colleagues at UCL, Lidia van Driel-Gesztelyi, led a team that created a series of models of the complex web of magnetic field in the corona. And, just like Parker, she took a dynamic approach and used the magnetic structures revealed by mathematical models to understand where and how the plasma trapped in the active regions might be re-routed and be able to escape.

The team found that the magnetic web around the equator can include 'null points'. Null points are interesting features and even though they are effectively regions that are devoid of magnetic field, magnetic fields can approach them and pass through. But as they pass through, they experience a change in where they are connected. This happens under special conditions in a process known as 'magnetic reconnection' and it has important consequences for the plasma. As the magnetic field reconnects, the plasma can move from being trapped in the magnetic field of an active region to being on 'open' magnetic field lines, which means the plasma is released into the slow solar wind flow, highlighting once again the dynamic nature of the solar corona.

This can be happening at several locations on the Sun at the same time. If you create a map of the composition of the plasma across the corona, as another of my colleagues, Dave Brooks, has done, you can look for the regions which show the same composition as the slow solar wind. In the images of the solar corona in ultraviolet light (see plates 8 and 9), you will see several sites in the corona which simultaneously show the right composition and upward-moving plasma which is able to get onto 'open' magnetic field lines to flow out and form the slow solar wind.

We are equipped with enough knowledge of the solar wind now to get back to the Voyagers on their epic mission. How far have they got? What have they found? But, first of all, what are our ideas about the outermost regions of the heliosphere? We need to think about this so that we can identify when the Voyagers are approaching their interstellar destination.

First, the solar wind will begin to slow down as it gets held back from expanding further by the interstellar medium: the tenuous collection of remnants of other stars' wind, nebulae and bits of dust. This sudden decrease in speed as the solar wind approaches the interstellar medium will be accompanied by an increase in density as the plasma is forced to pile up. After the turn of the millennium the position of this region went from a theoretical prediction to a measured quantity as the Voyager spacecraft finally reached it.

Voyager 1 reached the pile-up in 2004 when it was at a distance of 94 times the distance from the Sun to the Earth (94 'astronomical units', around 14 billion kilometres) and Voyager 2 was the runner-up in 2007 at a distance of 84 astronomical units (just over 12 billion kilometres). The two spacecraft were sent off out of the Solar System in different directions. This marked the first milestone on their way to interstellar space. The next milestone is to leave the flow of the solar wind altogether. NASA expected that this would happen ten to twenty years later.

Signs that Voyager 1 was coming tantalizingly close to passing out of the solar wind came fairly quickly though, in 2012. The excitement and anticipation led one American science writer and blogger to go online in October 2012, claiming that what he had seen in the data showed that the crossing had already happened.[*] From the outside, NASA looked like it had been beaten to this epochal announcement. The UK media responded with excitement and I was called for an interview on BBC Radio 4's *Today* programme. It was a Monday, so a quiet day for news, but it was a good opportunity to discuss how we would decide that the boundary had been passed: just how would we know that the crossing had been made?

The evidence needs to show that the material surrounding the

[*] http://blog.chron.com/sciguy/2012/10/more-evidence-that-voyager-has-exited-the-solar-system/.

spacecraft is no longer that of the solar wind but instead is that of the interstellar medium. The American journalist had pointed out in his online article that there had been a significant and persistent drop in the number of low-energy positively charged ions – particles that are found in the solar wind. He also pointed out that there had been a significant increase in high-energy atomic nuclei, or cosmic rays, from outside our Solar System that don't normally easily penetrate into the solar wind stream because the magnetic fields deflect them. He was making a good case that the crossing had indeed been made.

There is another measurement that Voyager 1 had been taking though – the magnetic field – and this indicated that the spacecraft could still be inside the heliosphere and detecting the magnetic field of the solar wind. Mixed messages like this show that defining the edge isn't easy. No wonder NASA were more hesitant to commit than the science writer was.

Even though predictions about the edge of the heliosphere have been made in the past, you always need data to find out what the real Universe is like. No one knew for certain if the 'edge' of the solar wind is a thin region or a thick one, for instance. Or whether particles are able to diffuse across the boundary, making it look like you are outside when in fact you are still really on the inside. The edge could be permeable to cosmic rays, for example. This is why NASA were waiting to make an announcement – they wanted to see what the magnetic field measurements were going to show in the days, months and possibly years ahead.

As we ponder what the heliosphere is like at the edge there are teams of scientists and engineers working on projects that will investigate the conditions close to the source. There is still so much more to learn about how the Sun creates and controls the heliosphere, which is a dynamic place. To do this, two spacecraft are currently being built. In America Solar Probe Plus is due for launch in 2018 on a trajectory that will take it closer to the Sun than any previous spacecraft. Solar Probe Plus will eventually spiral its way into an orbit just 6 million kilometres above the photosphere – 8.5 times the radius of the Sun and the closest we have ever been.

In Europe we are building a spacecraft called Solar Orbiter. Solar Orbiter is a European Space Agency mission that will work its way into an orbit that takes it closer to the Sun than the planet Mercury. It's an exciting mission – it will be able to hover over regions of the Sun for a few weeks at a time and in the later years of the mission its orbit will change and it will rise up so that it can start to look at the poles of the Sun. For many years now, scientists and engineers have been discussing how to take images of the Sun from up close as well as measure the solar wind directly around the spacecraft. We will be able to look at where the solar wind is coming from and then measure it when it gets to the spacecraft en route to the edge of the heliosphere.

I get very excited about Solar Orbiter but I'm a bit biased: my space lab at UCL is helping to build it. I can leave my office, walk down some stairs, through the common room, out to the workshops and see parts of the spacecraft being worked on right now. We are currently building sections of the Solar Wind Analyser, which will detect the elemental signature of the wind, and our electronics in the Extreme Ultraviolet Imager will mean this telescope can photograph the Sun in ultraviolet frequencies. It is hard not to get excited about a new spacecraft when bits are being put together in the building where you work!

It's also a technically challenging mission. All of the ten instruments it carries must work in the demanding environment so close to the Sun. The side of Solar Orbiter that faces the Sun will heat up to 600 degrees Celsius and will face the full onslaught of the charged particles coming from the Sun. But when Solar Orbiter and Solar Probe Plus are in place we will have two spacecraft that are exploring the creation of the solar wind and two exploring the edge, just in time before the latter's power sources run out. For the first time the beginning and end of the heliosphere will be studied simultaneously. I cannot imagine that this will ever happen again.

At the edge

A year after the excitement of August 2012, NASA scientists working on the Voyager 1 data had collected enough, and had had enough time to look for a consistent picture in the information coming back from the spacecraft, to confidently answer the question of whether it had indeed left the heliosphere. The particle and the magnetic-field measurements had been studied and interpreted. The scientists concluded that Voyager 1 had indeed left the heliosphere and entered a new, transition region, outside the heliosphere.

Upon reflection, the team saw that Voyager 1 had entered a new region of space on 25 August 2012. The strong increase in galactic cosmic rays suggested that it was in a region accessible by the interstellar medium. This day marks one of the major achievements of the twenty-first century as the first human-made object left the heliosphere, demonstrating our phenomenal drive to explore, no matter how hard the challenges are or how extreme the environment is.

Voyager 1 is now over 19 billion kilometres from the Sun* with this distance increasing by 17 kilometres every second. To send a signal to the spacecraft using radio waves that travel at the speed of light takes almost seventeen hours. A round trip is thirty-four hours. If you send a command to Voyager 1 at 9 a.m. in the morning, by the time it reaches the spacecraft and it responds you will not receive the reply until 7 p.m. the following night! A very drawn-out conversation. But, luckily, despite the

* As of October 2014.

astronomical distances, this conversation is still being had.

Voyager 1's leaving the heliosphere may not be the end of its story. In the back of the minds of some people was an even more significant event. In fact, a precedent had been set by two previous NASA missions, called Pioneer 10 and 11, which had been sent into the Solar System. They only visited the planets Jupiter and Saturn but they carried plaques on board that marked the time and place of their launch, just in case an advanced civilization should chance across them in the millennia ahead because they will also exit the heliosphere − although we won't know when as contact with these spacecraft is no longer possible. The two Voyagers carry a more sophisticated message. Engraved onto a 12-inch gold-plated copper record are images and sounds that act as a cosmic message in a bottle. The records contain information not just about where they were launched from but also about the humans who achieved this phenomenal feat and the world they inhabit. Even a needle is included so that the record, in theory, can be played.

Should extraterrestrials ever encounter the spacecraft and decipher images that are encoded on the record, among them they will find a set of images of the Sun, taken by the telescopes at one of Hale's observatories. Carl Sagan, who chaired the NASA committee that selected the contents of the disc, commented: 'The spacecraft will be encountered and the record played only if there are advanced spacefaring civilizations in interstellar space. But the launching of this bottle into the cosmic ocean says something very hopeful about life on this planet.' So, do the Voyagers have any hope of being found by another civilization?

Voyager 1 will not come across a star until around 40,000 years' time. And even then it's not a close pass. It will only get within 1.6 light years (15 trillion kilometres) of the star (named 'AC+79 3888') in the constellation of Camelopardalis. Once Voyager 2 has left the heliosphere it too will take 40,000 years before it comes within a couple of light years of a star. They probably have a very lonely future ahead.

The Voyagers are perhaps the most iconic of those early electronic explorers of the Solar System, for their longevity, for their scientific investigations and also because they represent a genuinely peaceful quest to learn more about the Universe – humans literally reaching out into the cosmos. The Voyagers weren't the first, though, and they certainly won't be the last. Many spacecraft have joined them. A small section of this fleet are a number that diligently and constantly observe the Sun – watching its every detail. These are the spacecraft that I use: the European/American SOHO mission, the Japanese/American/ British Hinode mission (the name translates as 'sunrise'), the twin NASA STEREO craft and NASA's Solar Dynamics Observatory. The space age opened our eyes to the real nature of our Sun and accelerated our scientific understanding of it. And all this ultimately came from technology that emerged from the Second World War and small groups of scientists and engineers who suspected that the Sun would look very different indeed when viewed from space.

10. Space Age

The sky is not the limit

Astronomers will go to great lengths for photons. The Mount Wilson Observatory, founded by Hale, was, as the name suggests, up a mountain, and even today is reached by a long and winding road. And modern telescopes in Hawaii and Chile are in places which, while great for a holiday, are certainly not easily accessible. This is all because putting a telescope up a mountain means that the photons coming from space have less atmosphere to pass through.

Photons carry valuable information and each and every one of them is important. But as they pass through the Earth's atmosphere, some of them are absorbed.* And the atmosphere is more damaging for some frequencies than others. It is not a coincidence (but rather an evolutionary convenience) that the frequencies which make it through our atmosphere, and are therefore the most abundant on the ground, are also the ones we use for vision. Telescopes on the ground are actually looking at the Sun through white-rose-tinted glasses; they are exploring the Sun by collecting visible light photons.

There are some frequencies of radiation near the visible part of the spectrum that also make it down to the ground. Enough of the ultraviolet, infrared and radio radiation emitted by the Sun makes it through the atmosphere for it to be observed with

---

* There is also the issue of 'seeing' – the bending of the light as it comes through the atmosphere, which degrades the image and causes stars to 'twinkle'.

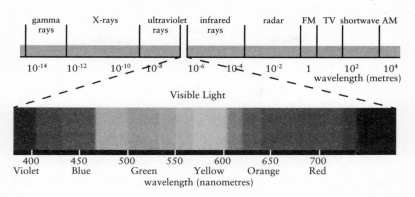

ELECTROMAGNETIC SPECTRUM

10.1 Schematic illustration of the components of the electromagnetic spectrum.

telescopes. But some frequencies are missing. No X-rays or gamma rays make it to the ground. Which raises the questions: does the Sun simply not produce X-rays and gamma rays, or does it produce them but the atmosphere is filtering them out? The only way to check is to take a peek at the Sun from above the atmosphere. Simple.

Or, rather: not simple at all. To get past enough of the Earth's atmosphere to make meaningful measurements of what is coming from the Sun requires getting to a height of at least 50 kilometres above the Earth's surface. There are certainly no mountains tall enough for this. Most passenger jumbo jets fly at only 12 kilometres above the ground. Even Concorde, with its maximum altitude of just over 18 kilometres, didn't make it. Fifty kilometres is a very long way up. Balloons could lift instruments to altitudes of up to around 40 kilometres, but beyond this was completely out of the reach of humans until the middle of the twentieth century, when rockets began to be used. The sudden introduction of rocket technology was motivated sadly not by curiosity about the Sun but rather by the Second World War.

The war was the catalyst for some revolutionary rocket research. The ambition for ballistic missiles that could deliver explosives across the borders of countries culminated in the development of a rocket-powered bomb: the V-2. The name stands for 'Vergeltungswaffe Zwei' – 'vengeance weapon two'.

The V-1 was its noisy jet-engine-powered predecessor. It was the buzzing of these engines that caused the people of England to name it the 'Doodlebug'. But while the V-1 may have been 'jet-powered' (using a pulse jet engine) it did not leave the Earth's atmosphere. The V-2 was the first human-made object to reach the upper limit of our atmosphere. Just under 1500 V-2 rockets were targeted at southern England during the war. And they came in silently, travelling much faster than the speed of sound. After the war, though, these rockets were repurposed for scientific use.

When it was clear that the end of the war was imminent, the German engineers who had led their country's rocket programme surrendered to the US. Almost 500 of them fled there. And this included the inventor of the V-2 rocket himself, Werner von Braun. Importantly, the US also got hold of around 100 rockets and immediately started launching them.

No civil space agency existed in America at that time so it was the job of the military to launch the rockets. The first launch took place on 16 April 1946, and as the rocket roared up from the White Sands Missile Range it reached an altitude of 6 kilometres. A good start to test the technology. It seemed that getting 50 kilometres up to look at the Sun would be possible. This was of interest to the Navy. Radio communications relied on favourable conditions in the atmosphere, conditions that were sometimes disrupted, with the Sun being the suspected culprit. Using a rocket could kill two birds with one stone: conditions in the atmosphere could be studied and astronomical observations made to see what role the Sun might play.

So scientists at the Naval Research Laboratory set out to use a rocket to answer a simple question: what does the solar spectrum look like at ultraviolet wavelengths when you get high up in the Earth's atmosphere? From the ground we can detect wavelengths ranging from around 300 to 1100 nanometres, which includes ultraviolet, visible and infrared radiation. The plan was to look for wavelengths shorter than any solar radiation that had been measured from the ground. This would reveal whether the Sun's radiation was affecting the upper atmosphere and might help shed some light on what was causing problems for radio communication. Scientists wanted to know because this would start to unravel the true range of radiation emitted by the Sun. Getting a detector to a high enough altitude, where the Earth's atmosphere is sufficiently thin and not able to absorb the photons, was the only way to find out.

On 10 October 1946 a rocket was launched carrying a simple detector made by the Naval Research Laboratory. It quickly broke through the stratosphere and into the very thin gas in the mesosphere. Reaching a final altitude of 55 kilometres it confirmed that the Sun emits more ultraviolet radiation than we are able to detect at the surface of the Earth.

This simple observation 55 kilometres above the Earth's surface was the birth of solar astronomy being done from space. The picture was not much to look at: a series of ultraviolet exposures from different altitudes. It was not even a picture of the Sun, just a measure of how much ultraviolet light it was producing. But it was the beginning of the scientific space age.

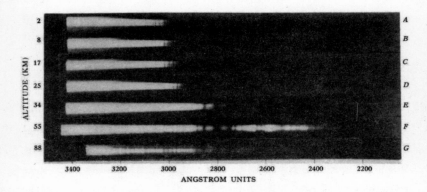

10.2 The ultraviolet spectrum detected during the 10 October 1946 rocket flight at different altitudes (© 1946, *American Physical Society*).

Rocket men

The UK space programme didn't begin as early as in the USA. There was no ready supply of rockets handed to the scientists but they had ideas on how to use them, and plans to get some were developing. Harrie Massey, a professor of physics at my own university, UCL, was keen to see the UK use rockets for research. He was also the Chairman of the Gassiot Committee at the Royal Society, which had been overseeing the Society's stance on studies of the atmosphere. UK scientists working in this area were well placed to adopt studies using rockets. The Gassiot Committee organized a conference with engineers and scientists from across the UK, the American rocketeers and the UK's Ministry of Supply, the Government department that was responsible for developing rockets so that the military could have ballistic missiles. But in the run-up to the conference things took an exciting turn.

On 13 May 1953, just as Massey was about to leave for the annual UCL staff–students cricket match, his phone rang. It was the Ministry of Supply calling to see if the committee he was chairman of was interested in using rockets for research. Massey said yes. The call reportedly made him late for the cricket match, but he would have had a good story to tell – the UK was going into space.

In 1955 the rocket that became known as 'Skylark' was approved, giving Britain its own national rocket programme to do science in space. With the funding in place, a sub-committee was set up at the Royal Society to advise on the programme. One of the founding members was Robert Boyd, who went on to establish the Mullard Space Science Laboratory at UCL, the department where I now work. The first Skylark launch took place from Woomera in Australia in 1957 carrying experiments to study the Earth's upper atmosphere. And several rockets quickly followed.

The big goal was to take a photo of the Sun from space. The first detectors were only able to confirm that the Sun was giving off short-wavelength radiation. Not long after the first V-2 launch, another rocket launched in 1948 showed that the Sun emits X-rays: a radiation of even shorter wavelengths than ultraviolet (it also used a very simple detector: a photographic film placed behind a filter made of beryllium that only lets through light with a wavelength of less than 0.4 nanometres). We didn't know what the Sun *looked* like in these wavelengths. But taking a photo is a challenging task because you need to have a stable camera to point at the Sun – very hard to do on a spinning rocket, but the US managed it. The first image of the Sun's X-ray faint glow was captured in 1960.

This image was taken when a rocket was launched carrying a set of pinhole cameras up to a height of 220 kilometres. They stared at the Sun for 286 seconds. They were delightfully simple devices – the best image was captured through a pinhole that was just over 0.1 millimetres across and created an image on the photographic film that was 1.6 millimetres across. The image is slightly blurred because the rocket rolled a bit, but the Sun was clearly visible. And the images showed that the Sun's X-rays are coming from the corona.

10.3 The 1960 X-ray image taken with a pinhole camera from a rocket. The rocket rolled and smeared out the X-ray features (© *Richard Blake*).

It wasn't just the US and the UK that were busy developing rockets and space technology. By the end of 1957 the Soviets had launched two satellites: Sputnik 1, which didn't do much more than beep (but analysis of its orbital path did reveal details of atmospheric conditions), and Sputnik 2, which barked – along with detectors for solar ultraviolet and X-ray emissions (and cosmic radiation) it also carried Laika the dog. Originally found as a stray dog on Moscow's streets, Laika sadly only survived a few hours in space. The US soon joined the satellite club with the successful launch of Explorer 1 in January 1958. And in 1958 the almighty NASA formed. The 'space race' was officially on.

The UK joined the orbital club on 26 April 1962 with the Ariel 1 satellite, built and launched in collaboration with NASA. NASA were keen to work with other countries and the UK was perfectly positioned after its studies into the Earth's upper atmosphere using rockets. Ariel 1 was designed to study how the Sun's high-energy radiation was affecting the Earth's upper atmosphere and carried instruments designed and built by scientists and engineers from the universities of Birmingham and Leicester, Imperial College London and, of course, UCL.

The team working on Ariel 1 included a young Ph.D. student

who was later to become important in my career. His name is Len Culhane and he went on to become director of the Mullard Space Science Laboratory, where I now work. He was also my Ph.D. supervisor and my first mentor in space science. Massey had given responsibility for the rocket research at the Mullard Space Science Laboratory to Robert Boyd, and Boyd was Culhane's Ph.D. supervisor. So Boyd's Ph.D. student became my Ph.D. supervisor. I see myself as the scientific granddaughter of the original UK space scientists!

The research that began to be done with rockets was immediately fruitful. This was discovery science – like reaching the South Pole for the first time or delving down into the Mariana Trench. The first instruments were in a totally new environment and showed what no one had previously ever set eyes on. Before Sputnik, about one major discovery about the Sun was made every year; after Sputnik, this rose to about three per year and the importance of space vehicles for understanding our Sun was made very clear. The Earth's atmosphere was no longer a barrier.

The highs and the lows

Perhaps one of the most ambitious space ventures is the story of Skylab – America's first space station – which was launched on the afternoon of 14 May 1973. By then NASA had been to the Moon and back, but there was no way for astronauts to stay in space for long periods of time. So NASA repurposed the Saturn V, a three-stage rocket built under von Braun's leadership to take humans to the Moon. But to place the astronauts in Earth orbit, the rocket had to reach an altitude of 435 kilometres, rather than traverse the 384,000 kilometres to our natural satellite, and this needed only two of the Saturn V's rocket stages. This freed the third to become accommodation and a workshop for the crew. This was the basis of the USA's first-ever space station: a cylinder 14.66 metres long and 6.7 metres in diameter.

Landing on the Moon had been NASA's big success in the 1960s, but the Sun became the focus in the 1970s as NASA intended to use the space station as an orbiting solar observatory. On the outside of the workshop was the Apollo Telescope Mount, which comprised eight instruments to provide solar images and spectra in the visible, ultraviolet and X-ray wavelengths. Skylab would provide a giant leap forward in our knowledge of the Sun. But the launch didn't go smoothly.

Heavy cloud cover obscured the view of the cameras monitoring the launch and the mission team didn't see that as the first stage fired, delivering the 7 million newtons of thrust needed for lift-off, part of it fell off! The vibrations shook free a shield crucial to protect Skylab from space debris, small pieces of rock and dust (micrometeoroids), and to shield the workshop from the searing heat of the Sun, like a thermal blanket to keep the heat out. Skylab was successfully delivered into orbit, but without the shield in place the temperature rose to 38 degrees Celsius within just a few hours. There was no way that the astronauts could visit Skylab in this condition. The problem delayed the visit of the first three-man crew, but when they arrived they had brought with them a solution to the problem: a space parasol.

This parasol was unfurled and fitted successfully onto Skylab, and two ninety-three-minute orbits around the Earth later the temperature had dropped to a comfortable level. On 28 May 1973 the workshop was ready to use. It's an interesting twist in the story to say that Owen Garriott was amongst the crew of three on that first mission to Skylab. Thirty-five years later his son, Richard Garriott, visited the International Space Station as a self-funded astronaut.

Across all the Apollo missions, NASA landed twelve humans

on the Moon, who stayed on the lunar surface for a total of 12 days 10 hours 35 minutes and 47 seconds. The first Skylab crew smashed this total by spending 28 days and 49 minutes in orbit, conducting science experiments and observations from their unique perspective. They had even been trained in solar physics to make sure the very best observations were taken. In total three crews, each involving three astronauts, visited Skylab and spent a total of 171 days, 13 hours and 14 minutes in orbit. They gathered images that showed the X-ray glow of the solar atmosphere in a totally new way. (See plate 10.)

Rather than fuzzy patches of X-ray emission, the corona was shown to be full of plasma loops of different sizes that arched up from the surface and back down again. There were dark regions that seemed to emit no X-rays at all – these became known as 'coronal holes'. There were sinuous, S-shaped features too. At these wavelengths the Sun doesn't look like a smooth ball of plasma at all. Seen in X-rays the corona is highly structured and has no clear edge. And it was realized that the X-ray emission was brightest above the strongest magnetic fields in the photosphere – above the sunspots.

The third and final crew left Skylab on 8 February 1974. Although plans were discussed that would allow Skylab to operate for many years and have its orbit boosted by a visit from the Space Shuttle that was then being developed by NASA, they never materialized. After the final crew had left and returned to Earth the engineers at mission control made their preparations to monitor and control Skylab's return home. With no orbital boost to keep the space station from being brought down by atmospheric drag, it began to lose altitude. Skylab slowly dropped as its orbit decayed and finally fell to Earth on 11 July 1979.

The re-entry was uncontrolled, although NASA engineers could make changes to the orientation of Skylab to try and influence the timing and location of the landing. Speculation and media interest in the fate of Skylab, which has a mass of almost 80,000 kilograms, soared. With the world watching, Skylab eventually came in over the Indian Ocean and scattered itself over the southern part of Western Australia.

The massive interest in Skylab led one American newspaper to offer a $10,000 reward for the first piece of debris to be delivered to them in San Francisco and NASA being issued with a littering fine by the town of Esperance, where some of the pieces fell. To this day there remain large pieces of Skylab in the local museum in Esperance. I made a pilgrimage there in 2011 to see what remains of the mission that transformed solar astronomy. (See plate 11.)

With the Apollo programme over and the phenomenal success of the Skylab mission under their belt, NASA launched their next solar observatory on 14 February 1980. This satellite was launched at the peak of solar cycle 21, giving it its name: the Solar Maximum Mission. Weighing more than 2,000 kilograms, it was about 4 metres tall and over 2 metres wide. Not designed to house humans, this satellite carried eight instruments into space that together could detect photons across the gamma ray, X-ray, ultraviolet and visible parts of the spectrum and included a coronagraph so that the extended corona could be seen.

The Solar Maximum Mission was launched during the era of the Space Shuttle and it had been designed with this in mind. It was the first satellite that could be captured and serviced in space by astronauts working in the cargo bay of the shuttle if something went wrong. And unfortunately there were problems with the satellite right from the start.

The satellite was launched successfully from Cape Canaveral and placed into an orbit 400 kilometres above the Earth. But three of the instruments began to have problems and this was quickly followed by a major setback when the satellite's orientation system failed: the satellite could no longer be controlled and pointed directly at the Sun. Just as with Skylab, some in-orbit repairs were needed. Despite having been designed to be visited by the shuttle it took three years of preparation and lobbying to NASA and Congress before a shuttle rescue mission was on its way. When the Challenger shuttle arrived in April 1984 history was made, as the Solar Maximum Mission became the first satellite to be repaired in space.

The repair was audacious and the satellite turned out to be a tricky object to catch. When Challenger reached the same orbit as the ailing satellite it began moving in unison with the satellite around the Earth. Now it was time for the astronaut George 'Pinky' Nelson (it seems 'Pinky' was a childhood nickname, nothing to do with his fingers) to leave the confines of the shuttle and use the Manned Maneuvering Unit to propel him towards the satellite and capture it. The Manned Maneuvering Unit was referred to as a flying armchair, an apt description, but it was an unsuccessful attempt and resulted in the satellite going into a spin that could have potentially put an end to any subsequent attempts at capture.

The skill of the engineers at the Goddard Space Flight Center in Maryland was tested as they wrestled to get control of the satellite again. Meanwhile, the astronauts were making plans to catch the satellite with the shuttle's Remote Manipulator Arm. With limited fuel on board, there were only two opportunities to use this approach. The first attempt was unsuccessful but the astronauts held their nerve, and no doubt the engineers and the scientists back at Goddard held their breath, and finally the Solar Maximum Mission satellite was captured. The satellite was successfully repaired and released back to duty on 10 April 1984. It went on to gather observations for five years until it fell back to Earth over the Indian Ocean.

Atmospheric physics

Thanks to all of these space missions, over only a couple of decades our view of the Sun was transformed. We knew that gamma rays were produced in the heart of the Sun and these photons gradually lost their energy on the way out to eventually be released as ultraviolet light, visible light and infrared radiation from the photosphere. But we are now confronted with X-rays and ultraviolet light being emitted from the plasma in the corona, above the photosphere. Something new was going on in the corona to produce this high-energy radiation. The emphasis in solar physics shifted – from the Sun itself to its atmosphere.

The corona is not producing this high-energy radiation through the same process taking place at the Sun's core: it does not have the density for nuclear fusion. Instead, the corona is a tenuous plasma: a wispy, thin plasma of electrons, hydrogen nuclei (protons) and helium nuclei with a smaller number of ions from the more massive elements such as calcium and iron. For this thin ion and electron soup to produce X-rays though, it must be extremely hot.

The X-ray glow of the corona reveals its temperature. The X-ray emission is being created by extremely fast-moving electrons, only possible in an extremely hot plasma. There are two ways that fast-moving electrons can produce X-rays: as they get deflected by positively charged particles and change direction releasing a photon, or when the electrons are captured by ions and then lose energy, radiating X-ray photons in the process. All of this points to the X-rays we see coming from the corona being created by a plasma that is exceedingly hot – reaching temperatures of around 2–4 million Kelvin. (See plate 12.) The space age revealed our Sun to have an atmosphere that is 300 times hotter than the photosphere below.

This had been suspected before the first rockets were launched, but it was a controversial speculation. Logically, the Sun should continue getting colder as you go out into the atmosphere: not suddenly get much hotter. But now there was concrete evidence that this was exactly what was happening. It was also happening very quickly in what became known as the 'transition region', where the temperature of the plasma rapidly rises from around 10,000 to almost 1 million Kelvin. This discovery presented a puzzle to understand, which physicists rushed to try and solve. And the key to the puzzle seemed to lie with the magnetic field that threads through the corona.

The space age forced us to shift our thinking about the Sun as merely a ball of plasma – well, everything up to and including the photosphere is a glorified ball of plasma: gravity keeps these

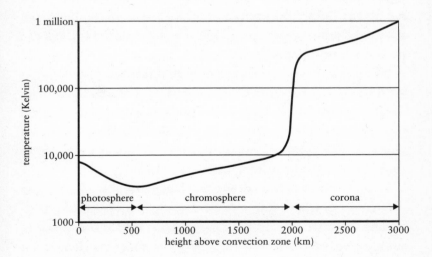

10.4 The temperature in the solar atmosphere initially decreases as you go up through the photosphere. But then the temperature starts to increase and this continues through the chromosphere and the corona. The rapid rise between the chromosphere and the corona is known as the transition region.

regions fairly ball-shaped and the regions have fairly well-defined edges. But things are much more complicated above the photosphere – forces other than gravity appear to be at play. In the chromosphere jets of plasma shoot up into the atmosphere and there is no sharp or well-defined boundary with the transition region. The same is true for the transition region's interface with the corona: the atmospheric layers are interwoven. And here we turn to the magnetic field.

Over the solar cycle the corona experiences an influx of magnetic field as more and more sunspots emerge into the photosphere. But the plasma of the corona is much thinner than that in the photosphere. In the photosphere there is enough plasma that it is able to push the magnetic fields around – not so in the corona. The amount of pressure exerted by a plasma is based on its temperature multiplied by its density, and the coronal plasma simply does not have enough density. Suddenly the balance of power switches and it is the magnetic field which can push the plasma about.

The photosphere in a sense acts as a boundary in two ways. As we have already seen, it is a visible boundary because it is where the light can suddenly escape from the Sun, giving it the defined edge we see in the sky. Secondly, it is the boundary where the magnetic field is rooted in and controlled by the plasma. Above the photosphere the plasma loses this control.

And there is a second major change in the relationship between plasma and magnetic field going from photosphere to corona. We have already met the magnetic field as the source of sunspots in the photosphere. There the magnetic field traps plasma and it cools to become dark. But in the corona the plasma trapped along the magnetic field structures is heated and appears bright in X-rays and extreme ultraviolet light. If you look at the photos of X-rays coming from the Sun, the bright areas match the massive loops of magnetic field.

The glowing super-hot plasma is trapped in the magnetic field and traces out its shapes. Magnetic fields that arch up from a positive polarity in the photosphere and back to the photosphere as a negative polarity form the arches seen in the space telescopes. But some magnetic fields pass through the photosphere and extend to such great heights that we don't see where they bend and turn back to the photosphere. The field lines continue straight out into the Solar System. These are the magnetic field structures we met before where the plasma rushes out to form the fast solar wind, leaving behind dark 'holes' in the corona. The magnetic field shapes everything.

Unsurprisingly, the current theory is that the magnetic fields are responsible for heating the corona, making it much hotter than the photosphere. Understanding how this happens has been a major question in solar physics in the intervening decades. Either the energy is directly extracted from the magnetic fields that thread the corona, or the energy is deposited by waves that propagate along the magnetic field. Either way, the magnetic field is responsible for the beautiful array of shapes seen in the X-ray glow of the plasma. And, either way, the transfer of energy from magnetic and wave to plasma heating is taking place in the chromosphere and the transition region.

One of the most recent missions to have been launched is NASA's Interface Region Imaging Spectrograph, most commonly referred to as IRIS. The satellite carries onboard instruments to study the chromosphere and the transition region with the aim of understanding how energy flows through these layers on its way from the photosphere to the corona. All the motions can be studied and the forms that the energy takes will be investigated.

Sending up spacecraft to study the Sun has been continuing ever since those first explorations. My work uses several different spacecraft, some carrying equipment and detectors designed and built in my department at UCL. Many people lament that NASA continued to launch rockets whereas the proto-UK space agency did not do the same. What people do not realize, though, is that the UK rocket programme lives on. Our Blue Streak missile became the first stage of a European rocket, which in later designs became the Ariane that is used today. And the UK was a founding member of what became the European Space Agency (ESA). Locally we played to our strengths within the funding landscape and specialized in the actual equipment that ends up in space. Our engineers built many instruments and our scientists analysed the data and made some incredible discoveries.

This is what we are going to look at next. From space we see the most dramatic phenomena in the Solar System. They are the result of the constant movement of the magnetic field in the photosphere and the emergence of new magnetic flux into the atmosphere which stores up colossal amounts of energy. But studying these phenomena has involved more than just rockets and spacecraft: there are balloons and atomic bombs as well.

11. The Flare Necessities

To catch a flare

Solar flares are the most powerful explosions in the Solar System. We saw a hint of their power through the story of Carrington and Hodgson, who in 1859 became the first humans to witness a flare. Any light source that can outshine the dazzling photosphere must be impressive in magnitude. And after the flare there was a major auroral display and disruption to telegraph lines, leading some to speculate that somehow this energy was able to propagate across the Solar System to us. Carrington and Hodgson saw an intense burst of white light coming from the Sun's photosphere. But now we know that what they saw was only the very base of the flare, merely its footprint, and that much more was going on above it.

In the century following Carrington and Hodgson's observation, the Sun could only be glimpsed in the visible light coming from the photosphere and chromosphere. When looking at the photosphere, seeing a flare was a rare event. On average, one 'white light' flare was seen every two years. But we started to work our way up the legs of solar flares when just the light of hydrogen alpha was used, which had been so successful for Hale and his work on sunspots. Looking through this narrow red wavelength of light, many more flares were revealed and an intriguing structure was seen above the bursts of white light.

Seen in this wavelength, flares show a bewildering array of shapes and sizes in plasma that is around 10,000 Kelvin, hotter than the plasma in the photosphere. Some flares are accompanied by sprays of material that shoot hundreds of thousands of

kilometres up into the Sun's atmosphere before falling back
again. Others are just small, compact flashes of light. There are
even some spectacular events that are accompanied by a sudden upward eruption of a vast amount of plasma into the
corona. Two things are clear from these observations: flares
involve a tremendous quantity of energy and there is more to
flares than the white-light flash. Before the space age, scientists
were like the metaphorical blind people, feeling the feet and
legs of an elephant and trying to guess what it looked like
further up.

It was time to look up at the rest of a solar flare. And that
involved not only a rocket, but a rocket suspended from a
balloon. A 'rockoon'.*

Guardians of the rockoon

In the summer of 1956, just before the peak of solar cycle 19, the
USS *Colonial* and the USS *Perkins* went out to sea on an unusual
voyage. The *Colonial* had been designed to transport vehicles and
launch them during amphibious assaults. The *Perkins* was a
destroyer, whose purpose was to escort and defend larger ships.
But that summer they set out on a peaceful mission. The USS
Colonial was being used as a floating lab by scientists at the Naval
Research Laboratory and the cargo it was carrying included a set
of rockets and balloons. The USS *Perkins* had been enlisted to
help. Together, they headed out into the Pacific to gather information about the most energetic explosions in the Solar System.
They were going to use their cargo to try and catch a solar flare.

There were several things coming together by the time the

---

* Not to be confused with 'Rocket Raccoon', which is less a scientific instrument and more of a Marvel Comics character. 'Rockoon' has nothing to do
with raccoons at all.

USS *Colonial* and USS *Perkins* sailed. The rocket technology that was being used to study the Sun from above the atmosphere was becoming established and there was a change in attitude about the radiation that might be emitted during a solar flare. The view that flares were a *chromospheric* phenomenon, shining only in visible light and confined to the lower atmosphere, was changing. A new view was emerging that flares might be a phenomenon that emits very high-energy radiation – like X-rays. It had been noticed that there were breakdowns in radio communications occurring during solar flares. This implied that solar flares were giving off enough X-rays to have a substantial impact on the Earth's ionosphere so that radio waves used for communication were no longer predictably bouncing off it, and instead were passing straight through.

But using a rocket to look at a solar flare is very difficult as it involves launching the rocket right as a flare is going off. Flares are unpredictable and fleeting events. And rockets only have a few minutes to make their observations. To get the brief observation window of a rocket to overlap with a short-lived solar flare means the rocket must be ready and able to fly at a moment's notice. The Naval Research Laboratory scientists solved this problem by using rockoons.

The word *rockoon* is not just a hybrid of the two words *rocket* and *balloon*, but also a blending of those two technologies. A balloon was inflated until it had around the same volume as a couple of double-decker buses, and was attached to a rocket, lifting it to around 20–25 kilometres. The rocket then waited, bobbing beneath the balloon until a flare went off. With very little delay the rocket was launched straight up through the balloon (this burst the balloon, and while not much thanks for its help, it did remove the problem of what to do with it afterwards). It was a novel approach. And there was something poetic about using helium balloons to help study the Sun.

Launching from a balloon meant the rocket could reach a much higher altitude than if it was launched from the ground. And it was simple and cheap and meant that launch towers weren't tied up with rockets that may or may not be launched – depending on whether a solar flare would happen. The balloon, with the rocket dangling underneath, would lift off from the USS *Colonial* and it was the job of the swifter USS *Perkins* to track the balloon as it drifted in the wind. Then, when word came through from a ground-based telescope that a flare was in progress, a radio signal was sent up to the rockoon and the rocket fired to head off and make observations with the detectors it carried on board.

The *Colonial* carried enough rockoons to launch one a day for ten days to try and catch a flare. Even though the Sun was close to solar maximum, when flares are most frequent, there were no flare sightings for the first three days. On the fourth day, two flares! But no rockoon had been lofted. On the fifth day there was success. A flare occurred, the rocket launched and, crucially, X-rays created by the flare were detected. This was the first step in answering the question about what was causing temporary changes to the ionosphere, but it opened a whole new kettle of fish. How were solar flares getting enough energy to create a brief but intense X-ray flash that could outshine the whole sun?

The problem was not even as simple as explaining how flares were producing such high-energy radiation: they were also producing unexpectedly low-energy radiation! It was discovered in 1942 that flares were producing long-wavelength radio waves, which are much lower energy than the visible spectrum or even infrared radiation. This was actually a serendipitous discovery, when a radio source was detected that overwhelmed signals received at a British radar station. It was soon realized that the radio waves swamping their receiver were not terrestrial: they had come from a flare on the Sun!

Any explanation scientists could come up with for flares had to include how they were such multi-wavelength events. They somehow produced all types of radiation seemingly at once.

Politics and plutonium

By the end of the 1950s enough data had been collected to show that bursts of X-rays were a fundamental part of flares. And since X-rays are emitted by incredibly hot plasma they showed that the solar flare had heated regions of the Sun's atmosphere to temperatures as high as 10 million Kelvin; this is several times the temperature of the ambient corona and a good percentage of the temperature in the Sun's core. Each observation using a rocket unveiled another part of the overall picture of what a solar flare is and it drove the early space pioneers to want to see more and for longer. They needed to observe the Sun for longer than a rocket or rockoon could achieve. They needed to use rockets to put solar observatories into orbit around the Earth.

The first solar observatories to be launched came once again from the Naval Research Laboratory. They began launching solar satellites in 1960 with the SOLRAD series (SOLar RADiation). They were small satellites, around half a metre across. NASA followed in the NRL's footsteps when it launched its Orbiting Solar Observatory on 7 March 1962. This satellite provided images that could pinpoint the location of the flare and took very detailed information about the ultraviolet and the X-ray spectra of flares.

The Soviets were working hard on understanding solar flares too and launched their missions Cosmos 166 and Cosmos 230 in 1967 and 1968, carrying telescopes that could see more detail in the corona. What was seen led to a change in perception of where to look to find out more about solar flares. The Soviet

observations revealed that even though scientists had been looking in the right *wavelengths* to study solar flares, they had been
thinking about the wrong part of the atmosphere.

The Soviet satellites showed that the X-ray emission from
solar flares could be seen as high up as 20,000 kilometres
above the chromosphere. And not only that: they hinted that
the high-altitude flash of the flare came first – flares initially
burst into view at high altitude and the chromosphere only
lights up as a secondary effect. Were the observations made
in visible light during the previous 100 years something of a
red herring?

The space age showed that to really see the true nature of a
solar flare, observations must be made from above the Earth's
atmosphere as well as from the ground. There is simply no
avoiding this. But working with rockets and spacecraft meant
that suddenly scientists had to deal with the world politics that
came with what was still very much a military technology.

I've asked my more senior colleagues many times about the
early days of space exploration. One story illustrates well the rush
to utilize all that space technology could offer, for military reasons
as well as scientific. It turns out that the primary aim of the SOL-
RAD spacecraft we met before was not to look at solar flares: each
SOLRAD launch concealed a second, secret satellite, which was
sneaked into orbit at the same time – a spy satellite called GRAB.
Designed to locate Soviet defence radar sites from 800 kilometres
above the Earth, the GRAB spacecraft were the first military
intelligence satellites.

And military operations were not just limited to being stow-
aways on scientific missions: sometimes they actively disrupted
them. The space age was born during the Cold War and there
were clashes as both scientists and the military planned to use the
space above the atmosphere. Things came to a head in 1962, when
the American military programme delivered a destructive blow to

the recently launched Orbiting Solar Observatory and Ariel 1 satellites which, due to the secrecy surrounding the military work, the solar scientists didn't see coming. The US military set off an atomic bomb in space.

In July 1962 the US Air Force launched a nuclear bomb from Johnston Island in the Pacific, around 1500 kilometres southwest of Hawaii. This was the Starfish Prime nuclear test: a 1.4 megaton thermonuclear bomb. It was actually one of four related tests, but the other three all failed. Two were aborted and blown up (in conventional, not nuclear, explosions) after launch but before they reached space, whereas the last one failed to launch and was destroyed on the launch pad, contaminating the island with plutonium.

Starfish Prime, however, was successfully launched, and the nuclear bomb detonated 400 kilometres above the Earth, well into what we consider 'space'. As well as generating heat and light, the explosion also created X-rays and gamma rays, and emitted a swarm of high-energy electrically charged particles. This was the purpose of the test. If enough high-energy electrons could be generated, they could form a belt around the Earth that might knock out the space technology of America's opponents. For the test though, they knocked out the technology of America and its allies – as the high-energy electrons rushed out they reached the OSO-1 and Ariel 1 satellites. The particles destroyed the detectors on Ariel 1, rendering the satellite useless. This was the end of the first British satellite. The Orbiting Solar Observatory fared better: it struggled through the electron bombardment, survived and went on to become a successful mission, spawning a new series, which finally ended in 1975. Just over a year after Starfish Prime, the US and the Soviet Union negotiated a treaty that banned nuclear-weapons testing in the atmosphere, in space or under water.

As a slightly related side note, a remarkable clash of military interests with solar scientists was the unfortunate end to the

Solwind satellite. Launched in 1979 to study the solar wind, by 1985 it had aged but was still semi-functional and providing data to scientists. That is, until President Reagan chose it for the US Air Force to use to test its anti-satellite technology. The scientists could do nothing as their spacecraft was used as target practice and blown into pieces (285 pieces, as far as could be counted).

The modern solar flare

The space age revealed flares to be multi-wavelength animals as they produce emissions across the electromagnetic spectrum, from radio waves with wavelengths of metres to gamma rays with wavelengths of less than one thousandth of a billionth of a metre. They also showed that flares reached much higher in the Sun's atmosphere than anyone had expected.

Flares begin with intense and short-lived bursts of radiation, lasting from seconds to minutes. Long-wavelength microwave and radio photons are seen coming from the corona, while lower down in the atmosphere X-rays appear in patches. All X-rays are by definition of a very short wavelength, but these are the shortest of the short: so-called 'hard' X-rays. Gamma rays are also sometimes produced at about the same time. Then bright ribbons of plasma are seen to glow at hydrogen alpha frequencies and, on rare occasions, there are bursts of visible light flashes coming from the photosphere. These visible light flashes had been the part of the flare emission that the Victorian amateur astronomers Carrington and Hodgson had seen. The series of events in a solar flare occur in almost exactly the reverse order to that in which they were discovered!

Actually, all of these first bits happen almost simultaneously and are over in a instant, in what is called the 'impulsive phase' of the flare. Then, structures that shine brightly in X-rays and extreme ultraviolet start to form in the corona above the locations where the hard X-rays, gamma rays and visible flashes were produced. In the most impressive flares these structures look like a row of massive arches, lined up side by side: like a slinky that has been half submerged in the surface of the Sun, only it is a slinky big enough for the Earth to fit inside many times over! (See plates 13 and 14.)

These massive solar slinky structures emit X-rays and extreme ultraviolet radiation which slowly fades. This is called the 'gradual' phase of the flare as it can take hours for everything to return to normal. Initially the plasma in the slinky must have a temperature of millions of degrees to produce the X-rays, but as it cools the frequency of the radiation goes down as well, ending with the slinky emitting light at hydrogen alpha frequencies. Then they disappear. They simply fade away.

Thanks to the space age, scientists had gone from just looking at the footprints of solar flares to seeing the body, the tail and the head. Finally, the anatomy of the beast was revealed. And they knew that all of the radiation, no matter which wavelength, is emitted from particles making up the plasma in the solar flare. So by looking at what particle processes could produce the radiation – and in the order in which they are observed – a full description of what is happening inside the flare can be deduced.

The initial radio emissions can be explained by electrons moving through a magnetic field. If electrons are accelerated in a magnetic field they start to spiral around the field lines and this produces radio waves. So a flare seems to start with a burst of fast-moving electrons, many of which are accelerated downwards from the corona. But things really get exciting when those electrons get to the bottom of the corona and something gets in their way.

Initially it is easy sailing for the electrons: the plasma is rather sparse in the corona – but it starts to become more and more dense the further down they get. The hard X-rays that appear in patches tell us that some of the electrons are colliding with protons in the chromosphere and that this is producing the X-rays. Patches of hard X-rays show us where this is happening, revealing where the magnetic legs of the flare have their feet in the lower atmosphere.

The sudden influx of electrons into the chromosphere also gives energy to the hydrogen atoms there. They then release this energy as photons at the wavelength of the hydrogen alpha line, creating the ribbons of hydrogen alpha emission that can be seen with any back-yard telescope that has a hydrogen alpha filter. In this deluge of electrons there may be some that have enough energy to make it all the way through the chromosphere to the photosphere. Those electrons can deposit their energy there instead, and if this happens they heat the plasma and create the visible light emission that Carrington and Hodgson so famously saw.

Meanwhile, the protons and ions in the coronal plasma are also being accelerated, just as the electrons were. They too spiral down the legs of the magnetic field but their relatively large mass and kinetic energy mean they manage to travel all the way down to the solar photosphere without being stopped. They're not as easily braked as the electrons were! But, even for them, the dense plasma of the photosphere is the end of the line.

When the accelerated protons and ions from the corona are finally stopped by the brick-wall-like photosphere, it is quite a show. They collide with the heavy atomic nuclei, such as carbon, nitrogen and oxygen, and these atoms then get rid of this sudden additional energy by admitting gamma rays. But some of the collisions with nuclei actually knock a few neutrons right off them. These liberated neutrons race off at high speeds, causing all sorts of new problems. When they finally slow down after a few collisions, they'll cause a proton to become deuterium (heavy hydrogen that has a proton and neutron in the nucleus), admitting more gamma rays. In all the high-energy excitement, even some antimatter can be produced! Antimatter particles have the same mass as their regular particle counterpart, but they have opposite properties, like an electric charge. Flares can produce anti-electrons ('positrons') from radioactive nuclei. With so much matter around they do not last long, quickly being annihilated in a collision with an electron and producing – you guessed it – more gamma rays.

All these collisions transfer energy from the fast-moving particles of the solar flare into the material they slam into and this means that the plasma has been greatly heated. The chromospheric plasma is suddenly heated from tens of thousands Kelvin to tens of millions Kelvin. And this rapid heating increases the plasma pressure in the chromosphere so that the second phase of the flare begins. Unable to shed the energy via radiation as quickly as it is being deposited by the charged particles, the heated plasma in the chromosphere expands and rises back up the magnetic field that the charged particles have just flown down.

This is what forms the slinky. The magnetic field lines the particles came down have now formed a row of massive arches. The super-heated plasma rises up and fills these arches and we can watch this happening with our space telescopes. Initially the plasma shines very brightly in soft X-rays and extreme ultraviolet light. But the plasma trapped by the magnetic field then cools as it radiates photons, and the X-ray and extreme ultraviolet emissions fade away, sometimes taking many hours to do so.

All the commotion with collisions in the chromosphere and photosphere has also done more than just cause the plasma to heat and rise up. It can also set off sudden bursts of sound like a hammer striking a drum. This burst of sound is called a 'sun-quake'! (See plate 15.)

We've seen from helioseismology that the Sun is always ring-ing like a bell from the constant movement of plasma causing sound waves inside it. A flare on some occasions is like hitting a softly ringing bell with a sledgehammer. These sunquakes were first discovered in 1998, when a very unusual pattern was reported in the plasma oscillations at the photosphere: a series of concentric circles. It was realized that these sound waves start in or close to the photosphere. They race into the Sun but are then bent upwards and appear at the photosphere in giant con-centric circles about the point where the hammer first hit.

But let's not get distracted by the drama of a flare. There is a mystery now to solve: how did so many electrons in the corona suddenly manage to be accelerated to such crazy speeds?

The elephant in the room

The fact that I find most staggering about solar flares is that to explain the amount of radiation we see, there must be one billion billion billion billion electrons accelerated, every second, to speeds approaching the speed of light. Something in the Sun is acting as an amazing particle accelerator. And it must be a very high-powered one.

The electrons carry in total one million billion billion joules of energy as they go crashing down from the corona. To put it in perspective, the 1.4 megaton Starfish Prime nuclear bomb released around 6 million billion joules of energy. This means a solar flare is millions of times more powerful! A single solar flare on the Sun is the same as 170 million Starfish Prime nuclear bombs all going off at once.

As we saw earlier with the work of Mayer and Joule, one of

the big scientific advances of the 1800s was the realization that energy cannot be created or destroyed. Energy can only be transformed from one form to another. So the energy that drives a solar flare must already be in the corona and somehow changes form to accelerate the electrons. But where is it hiding?

Both the Sun's core and the Starfish Prime nuclear bomb hide their energy in the same place: inside atoms. One of the big scientific advances of the twentieth century was realizing that energy could be released from within an atomic nucleus, in the case of the Sun, through the fusion process, fusing hydrogen into helium. In the case of Starfish Prime, the process is initiated through fission – the breaking apart of plutonium into smaller atoms. But fusion requires high densities and temperatures and fission requires very large radioactive atoms, none of which the corona has at the start of a flare. The energy must be hiding somewhere else.

There is one thing the corona does have in abundance though: magnetic field. And it turns out this is where the energy has been hiding all along. I've said before, I like to think of magnetic field lines like elastic bands. This is because it helps to visualize the magnetic field being twisted and distorted, forming the Sun's incredibly complicated magnetic field. Well, just like an elastic band, all that twisting and stretching causes energy to become stored in the magnetic field. And sometimes the field lines 'snap', releasing it all at once.

'Snap' is of course a gross over-simplification of a very complicated process. But it is not a bad analogy. We've all been pinged by a snapping elastic band releasing energy into us suddenly. Actually, an even better analogy would be if there were millions of stretched and tangled elastic bands that all snapped at once and then joined back together in a way that was much less stretched and tangled. Allow me to explain . . .

The normal situation in the corona is that the plasma is very thin, and so each charged particle can do whatever it wants, uninterrupted. And what charged particles like to do in a magnetic field is flow along magnetic field lines. So we generally assume that, in the corona, protons, electrons and ions can happily move about without getting in each other's way. Another way to think of this is that the corona is a very good conductor of electricity. It also means that charged particles in the corona flow along their own magnetic field lines and have no reason to jump across from one field line to another. They are loyal to their neighbourhood.

This is also the situation of the plasma and magnetic fields being 'frozen together' that we keep coming across. If the magnetic fields are made to move, they bring the plasma particles with them, and vice versa: if you force the plasma to move it will drag the magnetic field along with it.

However, under certain plasma conditions, particle collisions will disrupt the flow of the electric currents. Collisions within the plasma will send the particles in all directions and they get decoupled from their original magnetic field lines. This separating of particles and magnetic field lines means they can move independently of each other – they are no longer frozen together. As soon as the magnetic field lines have the freedom to start moving about independently of the charged particles, they can have the freedom to reconfigure themselves.

We conceptualize the reconfiguration as magnetic field lines being broken and rejoined as they diffuse through the plasma. But in reality they do not actually snap and break: it is just a great visual aid to portray what happens. The theory of this process has become known as magnetic reconnection and it is now in common use within the scientific community. I actually imagine magnetic reconnection as being like snipping two elastic bands and gluing them back together in a way that makes a new connection between them.

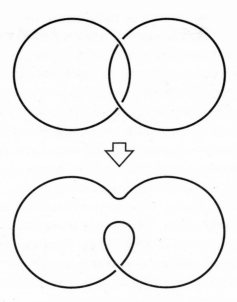

11.1 Conceptual reconnection: take two separate loops, break them apart at their top crossing and then join them together.

Despite our centuries of investigating magnetic fields on Earth, it was trying to understand solar flares which led to the development of the theory of magnetic reconnection. The reason magnetic reconnection is so important for solar flares is that the process converts energy that was stored in the magnetic field into energy of the motions of the particles. A vast amount of energy is liberated from the magnetic fields as they reconfigure and any electrically charged particles in the vicinity of the reconnection region are kicked into action and accelerated down from the corona.

Understanding that magnetic reconnection is the process which is at the heart of all solar flares means that every observation that has ever been recorded is actually of the secondary effects of the energy release. Really, we are inferring that reconnection is taking place by seeing the consequences of energy being transformed from magnetic into a form that we can observe which is the consequence of charged particles, particle interactions, nuclear process and plasma heating.

This theory that explains solar flares has been in development for decades. Indeed, it is still in development today although the foundations of an explanation are in place. Even though we have an overall understanding that magnetic reconnection is responsible for solar flares, the theory is not yet complete.

The problem is that the sites in the corona where the magnetic reconnection actually takes place are too small for us to see with current telescopes. The reconnection sites are probably less than 1 kilometre across, whereas our best coronal observations currently come from NASA's Solar Dynamics Observatory, which can resolve down to structures 700 kilometres in size. So we are not yet able to see directly into the reconnection region, where the important physics is happening.

Still to be tested are the theories of exactly how the charged particles are accelerated from the corona to the chromosphere and photosphere. How does the energy actually transfer from the reconnecting magnetic field to the particles in the plasma? Is this occurring in the vicinity of the reconnection region because of electric fields that act as a particle accelerator? Or maybe it is occurring in the reconnection region itself? Another possibility is that turbulent motions are established – irregular flows that stir up the magnetic field and plasma, leading to changes in the plasma density and magnetic field. Observations on very small scales are needed to investigate. It is as important as ever to develop new and better ways of observing the Sun.

And there is still more to be learnt about the way the energy is transported to the lower solar atmosphere. Acceleration of electrons from the corona to the chromosphere is a mechanism that works, but it is not without some mysteries.

The number of electrons needed to generate the observed

brightness of hard X-rays is much higher than the number of electrons that are in the vicinity of the reconnection region. To get the number needed, electrons need to be gathered from a vast volume of the corona. Either this is actually happening and our observations haven't yet been able to reveal it, or we need to develop better theories. And one of my colleagues at the University of Glasgow thinks the latter is what we should do.

Lyndsay Fletcher is a solar physicist working in a group at the University of Glasgow, where there is a long history of studying solar flares and developing the theory to explain them. The idea that the electrons that stream down from the corona lose their energy by colliding with particles in the chromosphere was developed there. She and her co-worker Hugh Hudson developed a complementary idea whereby the energy is carried to the chromosphere not just by particles, but also by waves that ripple along the magnetic field lines. This is an interesting idea as during magnetic reconnection the shapes that the magnetic field takes on change, and waves can be created as they snap back into place. Like letting go of a plucked guitar string, waves propagate along the magnetic field lines. In a guitar, the wave energy can be transformed into motion of the air and generate sound. On the Sun, the wave energy can be transformed into an acceleration of the charged particles.

Importantly, if true, this energy transformation takes place in the chromosphere, where the coronal electrons have accumulated after their journey from the corona. This theory means that fewer electrons need to be accelerated down from the corona because they can be re-accelerated in the chromosphere as the magnetic waves ripple through. It's a tantalizing idea. The hunt is now on to detect the waves and test the theory.

Back to what we do know: the last question is what causes the magnetic reconnection to start in the first place and trigger a solar flare? It could be that in the evolving magnetic field of the corona, magnetic fields get pushed up together and squeezed to breaking point. This produces a stand-alone flare. But often the situation is far more exciting.

I mentioned at the start of this chapter that the very early observations of solar flares showed that some of the plasma seen in the light of hydrogen alpha mysteriously started to move up as the flare went on beneath. This was initially undervalued by scientists and NASA, almost to their peril. It turns out that while solar flares may be the biggest *explosions* in the Solar System, above them are often the biggest *eruptions* . . .

12. Coronal Mass Ejections

'Coronal mass ejections' are the largest and most massive eruptions in the Solar System. The name describes an event when up to 10 billion tonnes of plasma blasts away from the Sun at millions of kilometres per hour. They can start relatively small – the size of a group of sunspots. But, as they leave, they expand to become many times larger than the Sun itself. They are beyond anything we can imagine. But, for something so awe-inspiring, they have been given a rather clunky, albeit descriptive, name. Yes, they are ejections of mass from the corona, but even the abbreviation 'CME' is nowhere near as catchy as their cousins: 'solar flares' and 'solar wind'. Something like 'solar eruptions' would have been better; I'd even settle for 'sun blasts'. But, no, CMEs it is, which makes them somewhat easy to disregard.

Putting nomenclature aside: a mystery even bigger than their name is how something that becomes so enormous went almost completely unnoticed until the 1970s – obscurity through implausibility, perhaps. With hindsight we can say that there had been hints though, and, for centuries, CMEs had been hiding in plain sight.

For example, if you look at sketches made during a total solar eclipse in 1860 you can see what is almost certainly a CME racing away from the Sun. Its circular shape makes it stand out next to the radial spokes that we call streamers. And once the spectroheliograph had been developed, hydrogen alpha light showed that the plasma in the corona did indeed form structures that sometimes erupted upwards. But it was assumed that this material never actually left the Sun.

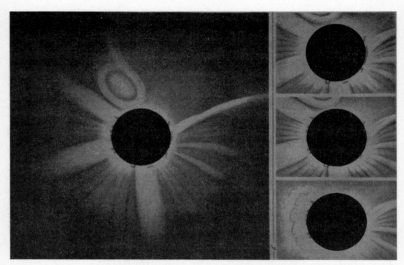

12.1 Drawing of the total solar eclipse of 1860 by Wilhelm Tempel (*Courtesy of the Royal Astronomical Society*).

It was so inconceivable that plasma could be ejected in this way and travel out into the solar system that when NASA were planning the Apollo missions to the Moon they did not take anything of the sort into account. They only worried about high-energy particles that were somehow linked with solar flares. They had no idea that as a coronal mass ejection ploughs its way through the Solar System it can create showers of high-energy particles too.

It was actually in 1971 that we opened our eyes and saw CMEs for the first time. On 13 and 14 December 1971, images were being taken by NASA's Orbiting Solar Observatory 7 which showed a section of the corona moving away from the Sun. Initially the technician thought there was an error with the camera, but it was soon realized that what they were seeing was a real event. In 1973 the head of solar physics, Richard Tousey, who had also recorded the Sun's ultraviolet spectrum from a rocket for the first time, published what they had seen and the CME age began.

This is the era when I entered solar physics and it has been an immensely revealing and exciting one. The old approach of using solar eclipses to watch the corona has found new purpose and we have been confronted with the challenge of finding out where the energy comes from to power these massive eruptions. Along the way we discovered that they told us something very special about the Sun's magnetic field. It is much more dynamic than we could ever have ever imagined and at times is like a pressure cooker, ready to explode.

Flying mountains

I said at the start of this chapter that CMEs involve a vast amount of material – as much as 10 billion tonnes of plasma. A big claim but one I can confidently make because we can calculate their mass thanks to the way in which we see them.

As we saw before, the corona is only seen in visible light during a total solar eclipse. With the photosphere blocked the corona is revealed because the electrons in the coronal plasma scatter the photospheric light. The corona is faintly visible because its electrons cause it to be effectively 'lit up' by the light from below. Some telescopes in space now have their own disc to replace the Moon to produce an 'artificial eclipse' but the process of viewing the corona is exactly the same. So when we see images of a CME, we are looking at light scattering off the electrons in its plasma. We know there must be protons and other particles along for the ride, but they do not reveal their presence when this technique is used.

We can, however, calculate how many there should be. The brightness of the CME from the scattered visible light gives us a way to work out how many electrons there are. So we just need to work out, for every electron, how many other particles there are as well. From that we can work out the CME's total mass.

Ignoring the trace elements that only appear in tiny quantities, roughly 10 per cent of the particles in the coronal plasma are helium and 90 per cent hydrogen. So for every helium particle, there are nine hydrogen particles. A hydrogen atom has one electron and helium has two, all of which have been stripped off in the plasma and are roaming free. So, our ratio of nine hydrogen for every helium will give us a total of eleven electrons. Helium nuclei are two protons with two neutrons, whereas hydrogen nuclei are a single proton. All said and done: for every eleven electrons we see, there must be eleven protons and two neutrons around too.

This is important to take into account because even though it is the electrons that are revealing themselves because they scatter photons, it is the protons and neutrons that give the CME its mass. The mass of an electron is nothing compared to these other particles. The mass of a proton is $1.6726 \times 10^{(-24)}$ g and the mass of a neutron is $1.6749 \times 10^{(-24)}$ g. The mass of the electron is about three orders of magnitude smaller than this. Sum all of this up and you find that there is an overwhelming amount of plasma in the ejection: somewhere between 10 million and 10 billion tonnes of material – roughly the same mass as Mount Everest.

You quickly see why CMEs were dismissed as unbelievable when you think about what it would take to blast something like Mount Everest off the Sun. This is only a few per cent of the mass lost from the Sun by the solar wind every day (the solar wind carries 1 million tonnes a second) and a negligible amount when compared to the mass of the Sun as a whole. But the huge gravitational pull of the Sun means that something really impressive is happening for a CME to be blasted away in one shot. I can't even imagine it being blasted off the Earth. But I can imagine something smaller being sent off our planet.

What would it take to get a fairly small object, such as a tennis ball, to leave the Earth and never fall back again? We know from our own everyday experiences that the faster we throw the ball, the higher it goes. Yet, no matter how hard we try, as the ball rises, it always slows down before falling back to Earth. But, in theory, could someone with superhuman strength throw the ball fast enough for it to escape the gravitational pull of the Earth? Thinking in terms of energy, the ball would need to be given enough kinetic energy, due to its motion, to exceed the gravitational potential energy the ball has at the surface of the Earth. We need to work out the ball's escape velocity.

A few equations can solve this for us. The kinetic energy* of the ball, indeed of any object, depends on its mass and the square of its speed. The gravitational potential† energy that the ball has is given by its own mass multiplied by the mass of the Earth and the gravitational constant – all divided by the distance the ball is from the centre of the Earth, the Earth's radius.

Like a good algebra problem, the mass of the ball appears in both equations and so we can cancel it out. What is important for finding the escape velocity is the mass of the Earth and the distance that the ball is from the centre of the Earth. It does not matter how massive the ball is. The same speed applies to any object, whether a tennis ball or a cannon ball, which is to be blasted off the planet.

$$\tfrac{1}{2}\, mv^2 = \frac{GMm}{r}$$

$$v = \sqrt{\frac{2\, GM}{r}}$$

That speed is around 40,000 kilometres per hour or 25,000 miles per hour (depending on how you like your units). Ignor-

---

* The equation for kinetic energy is ½ mv^2.

† The equation for gravitational potential energy is $\frac{GMm}{r}$ where G is the gravitational constant, M is the mass of the Earth, m is the mass of the ball and r is the Earth's radius.

ing any drag from the surrounding air, the ball would have to be travelling at 11.2 kilometres every second to escape, and we can follow the same logic to find the escape velocity for the Sun. We just need to scale up the size and the mass.

The Sun is enormous – beyond what we can imagine. We might have an idea of how large our local town or city is, and a vague sense of how big a country is, but the whole Earth is beyond what we can intuitively deal with. Well: over a million Earths would fit inside the Sun. Take a million of those things which are too big to understand, and that's how big the Sun is. And its mass is vast too, giving it a gravitational pull at the photosphere that is twenty-seven times as large as the force of gravity that we experience on terra firma.

Obviously there are even bigger problems than gravity that would make you not want to be on the photosphere of the Sun. But even if the Sun had a cool, habitable surface, 27g would not be fun. That 100 grammes of corn chips we had before would be pulled down as if it were 2.7 kilograms on Earth. A normal 75-kilogram human would weigh the equivalent of over 2 tonnes. Besides that, sustained gravitational force of 10g is fatal for humans, so 27g would not give you long to enjoy your holiday.

The mass of the Sun is roughly 333,000 times that of the Earth and its radius is 109 times greater. So scaling up our previous calculation accordingly gives the escape velocity of the Sun to be 618 kilometres per second. But mysteriously that is not always the speed we see CMEs moving at.

One of the very first things I did in solar research was to look at CME speeds. Curiously, there is a large variation – from 20 kilometres to over 2500 kilometres per second. If you look into these numbers you find out that 50 per cent of the eruptions leave the Sun at less than the escape speed. This simple statistic told me something important about coronal mass ejections – they don't need to reach escape velocity to escape. And this means they are not acting like a ballistic projectile. They aren't like a bullet leaving a gun. There must be another way that these ejections are lifted off the Sun.

To come back to the Earth analogy, another way that the ball could escape the Earth's gravitational pull is to be strapped onto a rocket so that the constant burning of the rocket fuel provides a continual force to overcome gravity and get into space. This is how the Saturn V rockets sent humans to the Moon and this is the approach that is relevant for coronal mass ejections: something must be continuously propelling them.

Well, 'propelling' does not do a CME justice. The Saturn V rockets were not able to send more than around 50 tonnes to the Moon (not including its own fuel). It takes a lot of energy to rocket something into space. CMEs not only have millions of times more mass, but they have the Sun's immense gravity to contend with. Launching them would require a phenomenal amount of energy.

And it's not hard to estimate just how much energy would be required. We know the mass and the velocity of CMEs from observations, so we can calculate their kinetic energy. Again, the kinetic energy is half the mass multiplied by the velocity squared. So, the kinetic energy of a CME carrying 100 million tonnes of material moving at 450 kilometres per second is of the order of 10 million billion billion joules. That is an insane amount of energy. That's 10 septillion joules. Or: 10 yottajoules. That's a yotta joules. Yet CMEs require at least that much energy to get them moving. This energy has to come from somewhere and this energetic mystery was the beginning of my career in solar physics.

Come fly with CME

These ejections captivated me when I started to learn about the

Sun and they pulled me in to want to understand more. By then, we had come a long way since the serendipitous CME discovery of the 1970s and there was now a spacecraft carrying a telescope dedicated to observing them. I was looking at images taken by coronagraphs carried on the SOHO spacecraft – artificial solar eclipse images made by placing a disc to block the light of the photosphere. But whereas a total eclipse gives a view of the corona for a few minutes, a coronagraph in space provides a view day after day and month after month: 14,000 CMEs were seen in the first three years of the SOHO mission alone.

Some of the ejections looked like amorphous blobs but some had a shape like a balloon with a bright centre. A particularly clear one with this shape was seen in the SOHO coronagraphs in 2000. Within the solar community it has become known as the 'light bulb' eruption because it had the shape of a glass bulb with a bright filament inside. (See plate 16.)

The SOHO coronagraphs are still being used today, despite the spacecraft having been launched in 1995. The SOHO coronagraphs take two to three images of the corona every hour, so they easily spot the eruptions and allow us to track them much further out from the Sun than had been possible with previous coronagraphs. These observations have been absolutely key to revealing how large these ejections become and that they are frequent events. There can be as many as six CMEs every day at solar maximum.

But despite the sudden wealth of observations of CMEs, we understood very little about them. I vividly remember being at a solar physics conference very early in my career and someone expressed frustration at how scientists had catalogued so many CMEs and could describe in detail what they looked like, but there was no detailed description of the physics actually causing them to happen in the first place. I decided that would be my first challenge as a solar physicist.

The main question about CMEs was the same question we had

about solar flares: where is the energy to drive them coming from?

Solar flares themselves were suggested as a possible energy source for CMEs because it was noticed that a lot of the time when a CME occurs, there is also a solar flare underneath. It was thought that the solar flares could somehow be blasting away the plasma above them. Solar flares release a vast amount of energy within minutes, which goes into heating the plasma in the corona to tens of millions of degrees. And the thermal energy, because of the plasma temperature, might provide the energy for the ejection and a large pressure that could expel the material.

But it was soon realized that there simply isn't enough thermal energy available in solar flares to do this. Moreover, flares and coronal mass ejections do not always occur together. Many flares have no ejections, and many ejections have no flares. We needed a mechanism to drive a coronal mass ejection that worked without an assisting solar flare.

One by one, possible energy sources were eliminated. Until the only one left was, once again, energy stored in the magnetic field.

I'm not sure which came first: my obsession with magnetic fields or my love of the mysteries behind CMEs, but they soon became intertwined for me. Work on solar flares had shown how energy could be released from magnetic fields through magnetic reconnection, as the magnetic field rearranged into a lower-energy state. A small region of reconnection could have large consequences and cause electrons and protons to rush down the magnetic field to produce all the aspects of solar flares we observe. Could magnetic energy be used to liberate something as massive as a CME?

The breakthrough is to realize that the mass in 'coronal mass ejection' is perhaps something of a distraction. Let's not focus on the mass for the moment. Actually a CME is an eruption of *magnetic field* escaping from the Sun that races into the Solar System. The mass just happens to be there for the ride. It is like looking at a tree moving about on a windy day and concluding it is pushing the air around. It's actually the air that is moving and the tree is merely being swept along with it.

CMEs are first and foremost the Sun's magnetic field evolving. But since the magnetic field and the plasma are frozen together, the plasma must also be swept upward. One of my colleagues, Tom Berger, is on a semi-serious campaign to rename CMEs as 'coronal magnetic eruptions'. I think he also likes the fact that Hale, discoverer of the Sun's magnetic field, used to refer to flares as 'magnetic eruptions'. Tom is perpetuating the phrase used by the first solar astrophysicist, albeit with a modern and accurate meaning. Which is not to downplay the significance of the plasma. But the hunt was now on for what could cause the magnetic field to suddenly erupt.

Second time around

The rapid changes in the magnetic field of the solar atmosphere, which lead to a vast bubble of it being ejected as a CME, reveal that there are times when it is highly dynamic. It has get-up-and-go on the most colossal scale and is in stark contrast to any magnetic field that we might interact with here on Earth. And this dynamic side needed to be explained. The solution came down to the very same magnetic fields we originally used to explain sunspots: ropes of magnetic field. In our explanation of sunspots, the flux ropes become buoyant and rise from the tachocline up to and through the photosphere. But to use these flux ropes to understand CMEs, we actually need to fix a problem I sneakily skipped over before.

Flux ropes do indeed form down at the tachocline because of the different rotational speeds within the sun stretching, amplifying and twisting the magnetic field. The plasma within them then becomes buoyant and they float up to the photosphere. At this point I said that they protrude up into the solar atmosphere and that where the magnetic field comes out of (and goes back into) the photosphere is where we get sunspots. All true, but the hard part is actually getting the flux ropes to push through the photosphere. The situation isn't as simple as I made it seem.

The less-dense plasma inside a flux rope makes it buoyant enough to rise up to the photosphere, but not buoyant enough to pass through it: the surrounding plasma changes too dramatically. Compared to the plasma inside the Sun, the plasma trapped inside a flux rope is light, but compared to the plasma in the atmosphere it's rather heavy. If you release a piece of wood under water it will float up to the surface because it is less dense than water. But the wood will not rise out of the water and float up through the air. It will bob around on the surface. This is what flux ropes do just under the photosphere.

This is all happening beneath the photosphere – out of our direct view. But this doesn't stop us believing that we have good reason to think this is what is happening – although it seems like a paradox: the photosphere stops the magnetic field, yet we see it breaking through to form sunspots. If you look at sunspot pairs as they form though, you see tiny pieces of magnetic field breaking through that gradually over time build up to form the spots. But it isn't the whole flux rope that emerges. Much like a floating piece of wood ends up half under and half above the water's surface, the flux ropes will end up partly above the surface and partly below. It is these half flux ropes that form the magnetic arches and arcades seen above the photosphere. But some (it's not known how much!) of the original flux rope always stays below the surface. (See plate 17.)

The reason why this partial emergence of the flux rope is disappointing for understanding CMEs is that if we could have a complete flux rope entirely above the photosphere, it could explain how CMEs are produced. Magnetic fields can hold energy if they are twisted, and so if there were a twisted flux rope, composed of a bundle of helical field lines, that had been bent into an arch, it could be bursting at the seams trying to spring back and straighten out the magnetic field. Flux ropes are also cohesive structures that can easily trap plasma. And so they would explain why CMEs are such a massive movement of plasma and magnetic field simultaneously: all that plasma could be trapped in the same flux rope.

We just need some way for flux ropes to form above the photosphere and be held there until suddenly released.

My interest was in working out whether magnetic reconnection in the photosphere could form twisted flux ropes that are tethered down by other magnetic arches. In short: a second-generation flux rope can form in the magnetic arches left over from the first-generation flux rope half poking through the solar surface. This idea was already around but needed testing against observations.

All that is left of the original flux rope is a series of arches of magnetic field that come out of the photosphere and curve around to come back down. But if two of those arches meet at their feet in (or close to) the photosphere, they can reconnect and form a helical field line. A small loop is also produced underneath the helical one, and this submerges below the photosphere. If a series of arches continue to meet and reconnect, a new flux rope will form as more arches reconnect into it. Importantly, there will be plenty of magnetic arches still reaching over this flux rope, holding it in place. At least, for a while.

Eventually, the energy in the twisted flux rope becomes too much and the magnetic arches over the top of it can no longer hold it in place. The upward force on the flux rope is caused by the magnetic field being more concentrated low down in the atmosphere than it is higher up, creating a steep gradient in the magnetic pressure. The magnetic arches succumb to the upward force of the flux

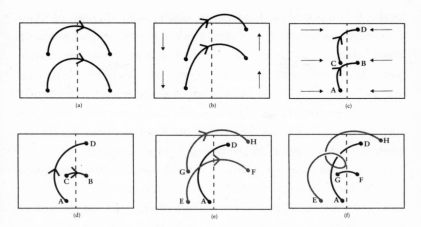

12.2 The creation of a flux rope from an arcade of magnetic field lines. Magnetic field reconnection happens between panels (c) and (d) between arch foot-points B and C. Magnetic reconnection happens again between panels (e) and (f) and this time involves foot-points F and G of another two arches (*Adapted from van Ballegooijen and Mastens (1989)*).

rope and are dragged into the Solar System. The flux rope breaks away from the Sun, resulting in the sub-escape-velocity speeds we observe – although some flux ropes involve so much energy that when they escape they move at phenomenal speeds.

This is all great in theory. But how do we know that flux ropes are definitely responsible for CMEs? We'd need to actually observe them in the corona. Frustratingly, it is very hard to see the magnetic field throughout the corona. The million-degree plasma is so hot it smears out any signature of the magnetic field that might be sent to us in the radiation. Magnetic fields are more readily measured in the relatively cool plasma of the photosphere.

So we need to use the age-old scientific test: see if our theory makes any predictions and then test those predictions. We need to find something uniquely 'flux-ropey' and see if there are signs of that in CMEs. Thankfully, the helical field lines in a flux rope gives us just such a property. Flux ropes can be twisted in two different directions: clockwise and anti-clockwise, which we actually call right- and left-handed directions of twist. And flux ropes with different directions of twist can behave slightly differently when they erupt.

Previously, solar scientists in America had noticed that the plasma in the corona would sometimes form an 'S' shape and that this indicated it was highly likely to erupt and produce a coronal mass ejection. And they had the idea that these 'S' shapes – which have become known as 'sigmoids' – are magnetic flux ropes. (See plate 18.) And, if they are flux ropes, the shape of a sigmoid would indicate the direction of the flux rope's twist. A normal 'S'-shaped flux rope would have a right-handed twist, whereas a mirror-image 'S' (Ƨ-shaped) would be left-handed. These details are needed to test the idea that sigmoids are the observational signature of flux ropes. And it is an exciting idea because if we can find flux ropes in the solar atmosphere, we might be able to solve the two main mysteries of CMEs at once: where does the energy come from and why does the magnetic field become so mobile?

Working with my colleagues, including Bernhard Kliem, who had worked on the theory of how flux ropes erupt, we began testing theoretical ideas. Firstly, that sigmoids form on the Sun as the result of magnetic reconnection between the arches of a partially emerged flux rope. Here we found many cases where the observations very nicely matched the theory – a good indication that these sigmoids are actually flux ropes! But there's more. The theoretical work made predictions about how flux ropes should behave. One example is that a flux rope which has a left-handed magnetic field twist should behave in a slightly different way when it erupts to a magnetic flux rope that is twisted in a right-handed sense. The curvature of the field lines of the flux rope means that a left-handed flux rope should, as it erupts, rotate in a counter-clockwise direction. A right-handed rope will rotate clockwise – albeit often only by a small amount.

So an erupting flux rope formed in an Ƨ-shaped sigmoid should rotate in a different direction to one formed in the normal S-shaped ones. We scoured observations and did a survey of which way the CMEs rotated when they erupted. Sure enough, the observations matched what the theory had predicted: the direction of twist in a sigmoid was related to how it rotated, supporting the interpretation that they are magnetic flux ropes.

There are still plenty of mysteries about CMEs and flux ropes for us to work on though. One is the exact role of magnetic reconnection. I said that in many CMEs flares occur as the eruption takes place. And flares are the result of magnetic reconnection. What's not clear is the order of these two events – our observations are not yet good enough to see whether the magnetic reconnection, and solar flare, actually begins before or after the CME erupts. It could be that the CME occurs first, causing the reconnection to take place. As always, there is much we still need to learn about the Sun.

The final twist

Coronal mass ejections go from being a curious feature of the corona to a vital part of the Sun's magnetic field evolution when you look at an obscure aspect of magnetic fields: helicity.

Helicity is a curious property of magnetism that isn't often talked about. It is a general measure of how distorted a magnetic field is. This distortion is ultimately responsible for the magnetic energy stored in the fields, but magnetic helicity is useful in its own right because it provides a way to measure exactly how distorted the magnetic fields are: how twisted, braided or interlinked. The plasma flows in the photosphere and below can lead to a high degree of distortion in the magnetic field of the Sun, and helicity provides a way to track how magnetic distortion moves through the Sun from the tachocline to the corona.

The concept of helicity actually goes right back to Karl Friedrich Gauss, the early-nineteenth-century mathematician, astronomer and physicist who contributed greatly to the theory of electromagnetism – the strength of a magnetic field is measured in units of 'gauss' in his honour. But the reason why magnetic helicity never became widely known about is that for

12.3 A coronal mass ejection imaged by SOHO showing a helical structure in the plasma. This shape is probably caused by a twisted magnetic field known as a flux rope (*Courtesy of ESA/NASA and the LASCO consortium*).

a long time it remained an abstract concept. When I started work, magnetic helicity was on the periphery of most solar physicists' attention. Today, it's a fairly commonly used phrase.

Helicity is incredibly important for solar physics for two reasons. Firstly, because the tangles in the Sun's magnetic field are so important, and, secondly, because magnetic helicity is such a tenacious property. Once magnetic field becomes tangled, there is no easy way for that magnetic helicity to be lost. We say that magnetic helicity is 'conserved'. Even during something as dramatic as a solar flare, there is the same amount of magnetic helicity after all that magnetic reconnection as there was before.

It can take a process as extreme as the rotation inside the Sun to produce a twisted and distorted magnetic field. One set of ideas suggests that the Sun's tachocline is a magnetic helicity generator, pumping out twisted magnetic field. But magnetic helicity can also be injected at other stages of a magnetic field's lifetime – by plasma flows in the convection zone, for example. Observations suggest that by the time the magnetic field emerges into the corona, it is twisted and distorted. And remember that this magnetic field keeps on emerging over the solar cycle.

My own interest in the role of magnetic helicity in CMEs came about when I read a paper written by a mathematician and solar physicist in the US called Boon Chye Low. 'BC', as he is known, had studied mathematics at the University of London but by the time of this paper had spent many years working at the High Altitude Observatory in Boulder, Colorado. The paper looked at the very fundamentals of coronal mass ejections. He thought about *why* the Sun might be producing such impressive ejections, which led on to *how* they might be happening.

BC wrote that coronal mass ejections were the way that the Sun could shed magnetic helicity and prevent it from being endlessly accumulated in the corona over each solar cycle because of the ongoing emergence of magnetic flux into the atmosphere.

Coronal mass ejections appeared to be a valve to remove magnetic energy that was stored in the twisted and distorted magnetic fields of the corona and couldn't be released any other way. It seems that coronal mass ejections are an essential part of the magnetic cycle of the Sun. In fact, BC wasn't the only person thinking along these lines. David Rust, who had put forward the idea that sigmoids were flux ropes, also suggested that CMEs were necessary for shedding magnetic helicity from the corona.

But just how much magnetic helicity is shed? Answering this was a major challenge. But in the early 2000s magnetic helicity went from being an abstract theoretical concept to being a quantity that could be worked out using observations. A small number of pioneering solar physicists developed ways to use observations of the Sun to follow magnetic helicity and I had the honour of working with two of them.

Pascal Démoulin works at the Paris Observatory, as he did when I first met him in 1999. As I know he will read this book, I won't say too much about my first memories of meeting him, but the mention of magnetic field caused his eyes to light up and sparkle through an otherwise dominating beard. Pascal specializes in using mathematics and computers to investigate the shapes solar magnetic fields could form. He had already modelled the magnetic configurations likely to lead to flares and was about to turn his mind to CMEs.

As we saw earlier, using the fact that magnetic fields leave their signature in the light coming from the Sun means that we can make a magnetic map of the solar surface. But this is only a thin slice through the Sun's magnetic field at the photosphere level; we cannot directly measure the magnetic field lines as they stretch up into the corona. Which is why Pascal's models are so important: he can take the magnetic data from the photosphere and mathematically simulate what it is likely to look like above that. Observations of plasma in the corona which trace out the magnetic structures can be used to check that his calculations are accurate.

Meanwhile, at UCL we had one of the world experts in helicity: Mitch Berger. The work of Mitch and Pascal meant we were able to calculate how much magnetic helicity there was in certain parts of the coronal magnetic field. Then we could take observations of CMEs and calculate how much magnetic helicity they were carrying away from the Sun. Which we could then match up to what we had seen on the Sun from the area they originated from.

As part of this work, we studied the magnetic fields in active regions of the Sun (identifiable by their sunspots) from the moment they emerged into the surface until the moment they dispersed completely. We watched as the once concentrated magnetic fields of the sunspots broke into smaller and smaller pieces, month by month, as the photospheric plasma flows spread them over a wider and wider area, until they were so fragmented that we couldn't track them any more.

This meant we could measure how much magnetic helicity was injected into the active region's magnetic field because of surface motions that move the magnetic field around and distort it, and how much was ejected because of CMEs, and then we checked whether the difference between these two values matched what was present in the corona.

The first surprising finding was just how many eruptions a decent-sized active region is capable of during its lifetime. One particular active region I studied produced thirty-five coronal mass ejections during the five months over which it was followed. There was a problem though. We can only ever see the side of the Sun facing us; we can never see the horribly misnamed 'dark side of the Sun'. Nowadays we have NASA's twin STEREO spacecraft, which are currently positioned to see the side of the Sun facing away from the Earth, but this was before they had been launched. So we estimated. Assuming that the region always produced coronal mass ejections at the same rate, we almost doubled the estimated number of ejections to sixty-five: over only five months – a phenomenal amount of activity. And a phenomenal amount of magnetic helicity shed into the Solar System. To think that before coronal mass ejections were known about, the corona was viewed as a slowly evolving environment where only large-scale magnetic structures varied as the solar cycle progressed from year to year!

This first surprising finding led to the second. The measurements we made showed that the active region's magnetic field was able to shed much more magnetic helicity in its CMEs than was being put in by the surface motions. There appeared to be a helicity reservoir inside the Sun that was somehow replenishing the stocks in the corona.

The discovery of coronal mass ejections was a milestone in developing our understanding of how the Sun works. But it also solved a long-standing question about what was responsible for causing some very intense changes to our own environment – our magnetic field.

For a long time it had seemed to scientists that the sunspot cycle and observations of flares correlated with magnetic changes here on Earth. But how could the Sun extend out and be responsible for these changes? Aside from electromagnetic radiation and the solar wind, there was no direct link between the Sun and the Earth. Then when coronal mass ejections were discovered, they offered a means to bring the Sun's dynamic magnetism out to the Earth.

And what began as an observation that these eruptions might be responsible for problems with the electric telegraph in the nineteenth century has become a realization that they pose a much bigger threat to twenty-first-century technology. In fact, they produce a whole new type of weather that we are now monitoring and forecasting: space weather.

13. Living in the Atmosphere of the Sun

Lessons from Apollo

Astronauts are the heroes of the modern age and meeting one is an honour. What they have achieved and what they have seen separates them out as exceptional people with exceptional stories. But there is one group among them who conjure up the deepest sense of awe: the Apollo astronauts who travelled to the Moon and back. They ventured across 384,000 kilometres of space to walk on the surface of the Moon, becoming the only humans to have ever set foot on another body in the Solar System. The Apollo astronauts are a key part of this story about the Sun because cutting-edge solar physics knowledge was needed to keep the astronauts safe. NASA knew that the Sun posed a threat to them at the times when it was active. And this same threat is now relevant to all of us today.

Very sadly, the first human on the Moon, Neil Armstrong, has since passed away. But the second lunar sightseer, Buzz Aldrin, is still alive and I have had the pleasure of meeting him. He was a guest on a BBC programme that I am a presenter on and, of course, both on and off air everyone wanted to talk to him about his experiences on Apollo 11. How did the Earth look from the Moon? What was the Moon landscape like? How does a Saturn V rocket launch feel? But, for me, meeting Buzz was the opportunity to ask something else: what did he see when he closed his eyes?

You see, Buzz and his fellow space passengers were some of the first humans to leave the protection of the Earth's atmosphere and go to the edge of its magnetic field, which together protect us from the dangers of space. For while space is almost as empty as the name implies, there are high-energy particles flying around: atomic nuclei, mostly individual protons, which race through the Solar System at speeds close to the speed of light. They were some of the first people to ever directly experience 'space weather' as the particles rained down on them.

These high-speed, high-energy particles were discovered in 1912 during a balloon flight that ascended to over 5 kilometres in height. Today we know that these 'cosmic rays' can be broadly split into two types: 'alien' and 'local'. The 'alien' rays have travelled vast distances before reaching the Earth, originally having been accelerated to very high speeds during supernova explosions far beyond our own Solar System. They are travellers from elsewhere in our massive Galaxy, the Milky Way. Other cosmic rays are far more 'local', coming as sporadic bursts that the Sun produces as a by-product of flares and CMEs.

Many cosmic ray particles of both types are either deflected by the magnetic field of the Sun or the Earth or else they collide with our atmosphere and disintegrate into smaller particles. The few that do make it down to the Earth's surface are not a health risk and generally pass by unnoticed. Only if you have a specially prepared tank of super-saturated alcohol vapour (known as a cloud chamber) will a cosmic ray make its presence known. As it moves through the vapour it will cause some droplets to condense out, tracing a ghostly cloud along its path. But mostly what you will see in a cloud chamber are the tracks left by the shower of those smaller particles, created when the cosmic rays reach the atmosphere. It's long been an ambition of mine to build a cloud chamber into my coffee table so that I am reminded that these particles are all around us.

On the way to the Moon, Buzz will have been exposed to cosmic rays flying in from all over the Galaxy. And there was a chance he could even have seen them directly. I had heard that if cosmic rays reach your eyes they could interact with your retina, causing you to see flashes of light. Thankfully I got a chance to ask Buzz about it, both live on the TV show and in more detail after the filming. My space lab is working on the next lander being sent to Mars (ESA's Exomars mission due for launch in 2018), a subject very dear to Buzz, so despite the rest of the cast and crew clamouring to talk to him as well, I was able to hold Aldrin's attention briefly by talking about future plans for the Red Planet.

And he did see cosmic rays with his own eyes! It was fascinating to hear Buzz describe the flashes of light that he had seen and thought were something inside the spacecraft. He saw them during their night and woke the next day to ask his crewmates if they saw the same. Mike Collins said no, Neil Armstrong said he saw a hundred or so. Buzz indicated that Neil Armstrong had a competitive streak! The next Apollo crew were told to look out for these flashes during the night and once their eyes had become dark-adjusted – they saw them too.

But NASA had known about this space radiation – even if they didn't know the Apollo astronauts would see it. The flashes of light were a novelty and make a good story now, but NASA knew that if the astronauts were to successfully land on the Moon and be returned in good health to the Earth, they needed to be protected from the particles. Cosmic rays are actually dangerous ionizing radiation and if the dose the astronauts received was high enough they could have been left with problems such as radiation sickness or even cancer.

What Buzz was seeing were probably the cosmic rays coming from the Galaxy. What were more worrying for NASA were the cosmic rays that come from the Sun. We call these local cosmic rays 'solar energetic particles'. They are the same kind of particle that Buzz experienced zipping through his eyes, but the numbers are very different. Buzz experienced a light drizzle, whereas a solar energetic-particle event is a torrential down-pour. And if an Apollo astronaut were to be caught in that downpour the consequences could be fatal.

So during the Apollo missions NASA had a network of tele-scopes monitoring the Sun and they had radiation experts working in the Mission Control Center Space Environment Console. They looked for solar flares because at that time flares were the only major solar activity that was known about. Coro-nal mass ejections and their role in creating energetic particles were still to be discovered.

The health risks that the particles posed to the small number of men going to the Moon, though, were weighed up against the political gains that the nation would achieve if they became the first to walk on the lunar surface. And solar energetic parti-cles were just one of the many risks they faced. So despite the Apollo missions being launched at the maximum phase of solar cycle 20 and on into the declining phase, when flares would have potentially been very frequent, the missions went ahead.

In hindsight the dose of high-energy particles received by the Apollo crews was small. And no solar energetic-particle event happened during any of the missions, so the astronauts never had to cope with the worst conditions that are possible. But plenty of particle events happened in between missions, such as between the penultimate mission and the last, the one that took Eugene Cernan to be the last man on the Moon. This is one of the largest particle events recorded during the space age. Looking back and knowing what we do today, it seems this would have given the astronauts moderate radiation sickness. It was sheer luck that there were no missions scheduled to take place then and there was, and still is, no way of forecasting these events.

The impact of solar activity isn't only of concern for the lucky few who have journeyed above the Earth's atmosphere. Today we know that the consequences are felt all the way down to the Earth, and even under its surface. And the main drivers of the worst space weather are not solar flares but coronal mass ejections, which hadn't even been discovered at the time of the Apollo programme. Things have moved on enormously and today making daily forecasts of our space weather has become an important aspect of modern society.

Victorian space weather

The Carrington event of 1859 continues to be our best example of an extreme space weather storm. It happened at a time when technology was advanced enough for some of its impact to be detected on the Earth but not so important to us that society was crippled. Recently, the Royal Academy of Engineering looked into the impact that a modern-day Carrington-sized event might have. It found that the effects could be significant but that above all we should aim to be prepared rather than be alarmed.

Waking up on the morning of a Carrington-esque solar storm, you will see no obvious warning signs of what is about to happen: the sunrise will look completely normal. Following the exact times on the day of the flare that Carrington and Hodgson witnessed, by 11.18 a.m. the flare will be in progress but nothing too severe has happened to the Earth yet. On the Sun, though, energy stored in the magnetic fields of a sunspot group is being released and converted into the kinetic energy of particles being accelerated down to the photosphere. There, the interaction of the particles with the dense plasma is rapidly heating it, causing it to radiate visible light photons. This visible light will take eight minutes to travel from the Sun to the Earth, as Carrington and Hodgson saw. If for some reason you happen to be looking at the Sun in white light, perhaps because you have made a trip to your local astronomical society, you may spot this burst of brightness, but otherwise there is nothing to notice.

Although it wasn't seen in 1859, the hot plasma rapidly expanded and rose up into the magnetic structure in the atmosphere above, glowing in X-rays and ultraviolet light. Today we would notice this with spacecraft looking at the Sun in these wavelengths. The only tip-off in 1859 was when these high-energy photons reached the Earth's upper atmosphere, ionizing the gases there and changing the Earth's electric currents and magnetic field. At the Kew magnetic observatory, a disturbance to the Earth's magnetic field was detected in their instruments. But this was only a tiny taste of what was to come.

As you now go about your normal day, a massive coronal mass ejection has blasted off the Sun, cutting its magnetic tethers and accelerating out into space. The coronal mass ejection has been launched from the sunspot group that also produced the flare, and is very near the centre of the Sun's disc. So the Earth is directly in the firing line.

The first blow comes from a cascade of high-energy protons of the kind that NASA was looking out for during the Apollo era. The fastest ejections can be travelling at speeds so high relative to the solar wind that they punch their way through, and that forms a shock wave. And in the shock, particles are accelerated to speeds approaching the speed of light, sending a shower of protons to the Earth. They reach the Earth in around twenty minutes. Not bad when you remember that light from the Sun takes just over eight minutes. There is no direct evidence that the Carrington event produced a shower of energetic particles, but modern data suggest that it would be unusual for an event of this size not to do so.

Following this initial commotion (albeit unnoticed by the general population) there will now be a few hours of relative calm. But then, just 17.5 hours after the flare would have been seen, the coronal mass ejection that has been racing our way at an average speed of 2300 kilometres per second will slam into the Earth's magnetic field. Now you notice that something is up.

When it arrived in 1859, the magnetometers at the Kew observatory went off the scale as a huge geomagnetic storm commenced – it remains one of the largest geomagnetic storms ever recorded. The storm created electric currents that flowed through the communications system of the era – the telegraph – allowing operators to switch off the batteries and operate using this natural electricity. The Carrington event was widely reported, but society at that time relied very little on technology and so relatively little disruption was caused. Were it to happen today, the story could be very different.

As you sit at home with a Carrington-style CME colliding with the Earth's magnetic field, what you absolutely don't want to see happen is for your power to go off. In the early hours of March 1989 a much slower-moving CME crashed into our magnetic field and Canada felt the effects when the Hydro-Québec electricity grid experienced strong voltage fluctuations. The fluctuations were so large that the grid's protection system was triggered, leading to the whole network shutting down in less than two minutes. The result was that several million people awoke to find that they had no electricity. It was a cold winter's morning and the power stayed off for more than nine hours. The lack of electricity to power society ended up costing the economy $6 billion.

This 1989 event was a wake-up call for just how much of a threat the Sun can pose to society as a whole. The Royal Academy of Engineering focused very much on the UK, where it found that a modern Carrington event could cause disruption to the electricity distribution in some regions, but probably only in remote areas, on the periphery of the network. Thankfully, since the scare of 1989, power grids are being hardened to be able to withstand this level of solar storm. Realistically, the power outage in the UK is likely to be along the same lines as that caused by other extreme weather events such as heavy snowfall. In these cases the electricity is normally back fairly quickly. Any countries that have not fully prepared might not be so fortunate though. Power grids around the world vary and some are better at withstanding space weather than others.

We'll assume that you are in a house that is on a CME-ready power grid. You may briefly lose power, but nothing too bad. Unfortunately, above your head, satellites are not so lucky. The Royal Academy of Engineering report found that perhaps up to 10 per cent of the entire satellite fleet might be disrupted, meaning that many prosaic aspects of our lives become impossible. On top of this, the changes to the ionosphere through which satellite signals pass mean that even some functioning satellites can no longer be relied on.

Satellite navigation signals will be significantly degraded. Banking activities may go awry as satellite navigation signals are often used to time-stamp money transfers. Airlines may choose to ground their fleets as changes to the ionosphere lead to the loss of radio communications and poor satellite navigation. And, on top of all that, the charged particles threaten the function of the microelectronics at the heart of the plane and may subject humans to a radiation dose as high as three chest CT scans.

There is one silver lining to all of this: the aurorae will be spectacular. The Northern and Southern Lights normally only occur close to the magnetic poles, but during a major geomagnetic storm they move further towards the equator. On the night of 1 September 1859 the northern aurora pulsated with a blood-red colour and was so bright that people in England were able to read their newspapers without any additional lighting. So your satellite TV has stopped working, your flight has been cancelled and you may not even have power at home: but at least you will get to witness one of the greatest astronomical displays ever.

Delivering a blow

In short: a CME is an ejection of solar magnetic field, and if it hits Earth in just the right way, it can really mess with our magnetic field. And that means that we feel its effects all the way down to the surface of the Earth. This has really made the study of the Sun far more practical than physicists could have ever expected. So far we have seen our understanding of the Sun being driven by curiosity, through serendipitous discoveries and deliberate attempts to answer questions. Piece by piece, we have put together a jigsaw that links the physical processes which explain our Sun and its control over the Solar System. Now we need to re-evaluate this jigsaw to understand possible threats from the Sun, using everything we have learnt from 400 years of science, over 100 years of studying the magnetic nature of our Sun and almost sixty years of observing from space.

First off: the impact that we feel at the Earth's surface is not the whole story. Above our atmosphere but within our magnetic field are ring-doughnut-shaped regions where electrically charged particles are trapped. These are the Van Allen belts, named after James Van Allen at the University of Iowa. Van Allen was an early pioneer of the space age and had used the V-2 rockets captured by the Americans after the Second World War for cosmic ray studies. He even worked with the creator of the V-2, von Braun, on Explorer I – America's first satellite: von Braun worked on the rocket and Van Allen on the scientific instrumentation. We have met other projects that Van Allen worked on already in this book as he was also involved in the rockoons. The detectors that he designed and which were launched on Explorer I in 1958 discovered the trapped particles. It was a game changer for the way we viewed the space above our atmosphere – it is not empty – and led Ernie Ray, a colleague of Van Allen, to exclaim: 'My God, space is radioactive!'

The radiation belts are an area of study today because many of our satellites orbit through them. The satellites that you use for communications and TV broadcasts all risk being damaged by the particles trapped in the belts. The particles can degrade solar panels and cause electrical sparks in the spacecraft that can damage the electronics. This is the same effect that we saw the US military trying to exploit before.

It was the Van Allen belts that the 1962 United States Air Force Starfish nuclear explosion was designed to influence. The US military were testing whether enough high-energy electrons could be injected into the radiation belts, making them more deadly and scuppering Soviet surveillance and missile technology. We know that at a minimum the electrons damaged research satellites. And by 1969, when the first Apollo landing occurred and Buzz Aldrin saw the flashes of light in his eyes, the electrons had diffused away and their levels had dropped to less than 10 per cent of the 1962 value. NASA were concerned about the Van Allen radiation belts but it turned out that the astronauts passed through with no problems.

So, it's not a perfect vacuum above our atmosphere (although it is better than any vacuum we can make on the Earth). The particles trapped in the belts come from two places: again local and far away. Some particles are stripped off from the top of our atmosphere and others leak in from the solar wind. It is unknown what the actual ratio of Earth-origin to Sun-origin particles is in the Van Allen belt. What we do know is that changes to the Earth's magnetic field that are the result of space weather can cause these particles to be accelerated down into the atmosphere and make the gases there glow. This forms the aurora, and the more particles that are accelerated, and the faster they move, the more dazzling the aurora displays. If that sounds familiar, it is a very similar process to what causes solar flares on the Sun – fast-moving electrically charged particles giving up their energy to gases, which makes the gas glow.

The global effect that a CME has on our magnetic field actually depends on the details of the CME's magnetic field. When the orientation of the magnetic field in the coronal mass ejection is opposite to the orientation of the field of the Earth's magnetic field, it creates an opening in our normally protective magnetic bubble. This changes the strength and shape of the magnetic field – affecting it all the way down to the surface of the Earth. The physical process behind this is also the same as the process that is at the heart of solar flares: magnetic reconnection.

As always, magnetic reconnection is best visualized as being the breaking and rejoining of magnetic field lines. In this case it joins up the magnetic field within the coronal mass ejection to the magnetic field lines in the Earth's magnetic field's outer layer, creating field lines that can be traced from the coronal mass ejection all the way down to regions close to the magnetic poles of the Earth. But the CME is still on the move, sweeping over the Earth. By the time a CME hits our planet it has expanded to be many times the size of the Earth. It does not so much hit the Earth as flood around it.

This means that as a CME flows around the Earth, it connects to the magnetic field on the 'front' of the planet (the Sun-facing side) and drags that magnetic field around the Earth to the far side. Even without a CME, the solar wind causes the Earth's magnetic field to deform like a kind of solar windsock on the night side. But the CME exaggerates this to a huge extent, so much so that magnetic field and energy are stored up in this region and magnetic reconnection can then occur in what's called the tail. This reconfigures the magnetic field and actually allows it to snap back towards the Earth and move back to the dayside, where it can reconnect once more with the magnetic field of the same CME and the whole process starts again.

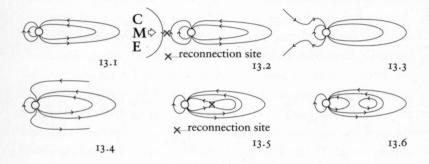

This circulation is known as the Dungey cycle (see figures 13.1–13.6), after the British physicist James Dungey who developed this theory whilst working at Imperial College London. Apparently the idea came to him during a eureka moment as he stirred his coffee at a street café in Paris – a humble origin for a theory that went on to be the backbone of our understanding of the way the Sun interacts with the Earth. The huge changes to the Earth's magnetic field create geomagnetic storms.

The impact on the Earth's magnetic field is strongest when a fast coronal mass ejection comes our way – such ejections deliver more of a punch. Our modern view of the Sun has shown us that all three types of activity – flares, coronal mass ejections and energetic particles – can be initiated at roughly the same time, but only in the worst cases (events which are fairly rare but which can happen at any time) do all three occur together. Which is what happened in 2012, when a CME similar to the Carrington one was actually launched by the Sun but, thankfully, we were able to watch as it missed the Earth and sailed harmlessly past us. In the coming decades we may not be so lucky. We had better be prepared.

Space weather

Today, monitoring of the Sun in the US is carried out by the Space Weather Prediction Center, in Boulder, Colorado. This is the latest incarnation of the Space Disturbance Forecast Center that was set up during the Apollo era, and which itself had a heritage that dated back to 1903, when interest in space weather was first aroused because of its effects on radio propagation.

In the UK, space weather forecasting is the responsibility of the Met Office: you can get your weather and your space weather information from the same place – very convenient. But to be able to make accurate space weather forecasts we need to collect the right data. Just as weather forecasts require weather satellites to make observations of wind patterns and the temperature of the oceans, we need space weather satellites to measure the patterns in the solar wind and the formation of sunspots that might produce flares and coronal mass ejections. If we have enough warning that a storm is coming, we can batten down the hatches.

So we are looking at what a space weather satellite should be like in order to forecast space weather in an accurate and reliable way. Up until now the vast majority of our data has been coming from satellites that have been designed for scientific research. The satellites I use are designed to understand the physics underlying phenomena like coronal mass ejections. This is important for understanding space weather, but they aren't designed to forecast it. What should a space weather monitor be like?

Well, it would need to be able to measure the strength of the solar wind, including coronal mass ejections, and the direction of the magnetic field before it arrives at the Earth. Actually, we already have some instruments that do this at an incredibly useful point in the Solar System to put a space weather monitor: 1.6 million kilometres closer to the Sun than the Earth is, known scientifically as the 'first Lagrange point'. The first Lagrange point is the location where the gravitational pull of the Sun combined with the gravitational pull of the Earth means it's possible to place spacecraft into orbit at that location. The spacecraft will always stay between the Sun and the Earth. And from this position it tells us what is coming before it arrives, allowing us to make a forecast of whether a geomagnetic storm might form. The drawback is we only have a warning time of about an hour for the solar wind and twenty minutes for a fast coronal mass ejection. But it doesn't have to be this way.

If we can get our satellites closer to the Sun, and make our measurements from there, we will have more time to make our forecast. One idea that I like has become known as the 'space weather diamond', so called because it makes use of four satellites that form a diamond pattern. These satellites are in orbit around the Sun, but in a way that from our perspective makes them seem to dance around the Earth, ten times further away from us than our current solar spacecraft, at the first Lagrange point, giving us ten times the warning time. Four satellites are needed so that, as they move in relation to the Earth, there is always one on the sunward side. This could be a costly mission to launch though, so solutions with fewer spacecraft are also being considered.

Another option is to watch the Sun and look for times when a coronal mass ejection is launched and heading towards us. Then we have around one to four days to make our forecast about whether it will produce stormy space weather. And if we want to look for coronal mass ejections that are headed our way, a good approach is to step to one side and look back at the space between the Sun and us. This approach has already been shown to be successful by NASA's STEREO spacecraft — twins that were put into orbit around the Sun.

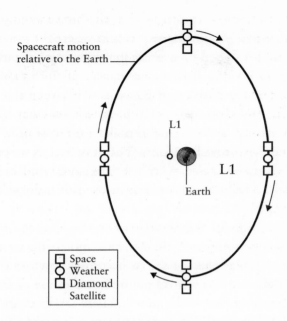

13.7 A schematic showing how four satellites forming a space weather 'diamond' could provide an early warning of emissions from the Sun that are headed our way. 'L1' indicates the location of the first Lagrange print.

One STEREO spacecraft is slightly closer in to the Sun than us so that it moves more quickly and drifts ahead of the Earth orbiting around the Sun; the other is in an orbit slightly further out so that it moves more slowly, causing it to lag behind. Gradually, the position of these spacecraft changed and they could both look back at the Earth and the Sun and the space in between. Both STEREO spacecraft carry a cleverly designed telescope that blocks the light of the Sun to enable us to track the very faint coronal mass ejections far out into the Solar System and predict which ones will hit us and when. Whilst the STEREO spacecraft have been useful to demonstrate a proof of concept, a drifting spacecraft is no good for space weather forecasting (they are currently behind the Sun), but luckily there is another option that we can make use of.

It is possible to place a spacecraft in orbit at a point 150 million kilometres behind the Earth in its journey about the Sun – the same distance away that the Sun is from us. This is the fifth Lagrange point. Looking at the Sun from this position allows us to watch the CMEs headed towards us but also peek over the edge of the Sun and see regions that are hidden from us on Earth. We can see what is about to rotate to face us. If we watch the surface and the atmosphere of the Sun from this viewpoint we can monitor the Sun's magnetic field and possibly get several days' more warning of stormy weather ahead. A longer forecasting time, giving us enough time to make our preparations, is the holy grail of space weather forecasting – like predicting the strength and motion of a hurricane as it is developing over the ocean. It's also an important place to take measurements of the solar wind directly, which is absolutely vital for making an accurate space weather forecast. The forecasts are made using models, and models need real data to drive them. We have to know what the real Solar System is doing.

But it's not only observations from space that matter. Novel space weather stations are being set up around the globe. And they can be much, much cheaper. One of my colleagues in Ireland has taken this approach. He has set up a space weather ground station on the site of what was once the world's largest astronomical telescope, a telescope so large that the locals named it the Leviathan. The Leviathan was built by the third Earl of Rosse. When it was completed in 1845 it became the largest telescope in the world, a title it kept for over seventy years. Now Peter is reviving this site by setting up a space weather monitoring station. He has persuaded the current Earl of Rosse to donate a disused building on the site and Peter's team has transformed it into the Rosse Solar-Terrestrial Observatory. From there they monitor the Sun's flare activity and look for changes in the Earth's magnetic field caused by the flares and CMEs.

It's not just forecasting that is important though. The reason why the Royal Academy of Engineering undertook the report is so that we can find engineering solutions to prevent the problems in the first place. We can improve our technology and make it more resilient to space weather, such as designing buildings that are to be constructed along fault lines to be able to withstand earthquakes. In the space weather context this might include changes to how power grids are designed and operated, or designing satellites with enough shielding to protect them from high-energy particles.

For me, one of the most important outcomes of the space age is that we have redefined our relationship with our local star. Space weather has brought a new relevance to understanding the Sun and an imperative to watching its behaviour. And behind the scenes there are teams of scientists developing the theories to explain what is happening to forecasters working to help keep the lights on, and across the country engineers are working to develop new technologies that won't be so susceptible to space weather.

Levitation

We started the chapter with the Apollo astronauts because they were some of the early adopters of space weather forecasts. From that point on it may have been felt that space weather is a rather sombre subject because of the problems it can cause for the technology that we rely on. A gloomy subject except for the beautiful aurorae that space weather creates. I want to end on a more positive note, one which reminds us that space weather is really about us understanding our place in the Solar System and the scientific processes that are happening all around us. If you allow me a final trip back to the Moon, we'll see that the science of space weather solves the mystery of why the Apollo astronauts saw something at sunrise that no one was expecting.

When the Apollo 17 command module went into orbit around the Moon in 1972 the astronauts had a clear view towards the horizon as they flew through the lunar night. Ahead of them, the Sun was just about to rise. Module Commander Eugene Cernan was ready to draw the view of the lunar sunrise and, as he watched, a faint bulging glow began to appear over the edge of the Moon. The glow wasn't the appearance of the Sun itself, but sunlight being scattered over the horizon by dust, both dust in the Sun's atmosphere and dust between the planets that has been laid down by comets and asteroids over the millennia. This glow was expected.

But there were some aspects to the glow that were very odd. There was a glow that spread out along the horizon and shafts of light were seen shooting upwards. The shafts looked similar to the shafts of light we sometimes see coming down through holes in layers of cloud on a sunny day. On the Earth these shafts are easy to understand because we have an atmosphere, and sunlight scatters off particles in it. On the airless Moon, the shafts of light were a complete puzzle.

The Moon simply isn't massive enough to hold on to any gas to form an atmosphere of its own and with no atmosphere there should be no suspended dust particles to scatter the sunlight. Yet somehow the Moon has dust fountains that give it a very tenuous atmosphere. These dust fountains need a way to be lifted up from the lunar surface. And with no wind on the Moon, the answer was in the charged particles trapped within the Earth's magnetic field.

The Apollo astronauts saw the dust fountains as they flew along their orbit and crossed over from lunar night to lunar day.

Just as on the Earth, on the Moon the line that divides day and night is called the terminator. It is along the terminator that the dust was seen to be lifted up.

The idea is that the dust is levitating because it is electrically charged. On the dayside of the Moon photons of sunlight that fall on the dusty lunar surface have enough energy to knock electrons out of the atoms making up the dust grains. Without a complete set of electrons the dust becomes positively charged. On the nightside there is no sunlight, though, and the question as to how the dust becomes charged is harder to answer. But this is where the nature of the space environment around the Moon comes into play. The dust elucidates the invisible.

The dust on the nightside could be given an electrical charge if the space around the Moon has electrically charged particles – electrons, for example. Electrons could attach themselves to the dust, making the grains negatively charged.

Along the terminator there will be a sharp change between dust that has a positive charge and dust that has a negative electrical charge and this creates an electric field. The electrically charged dust particles will feel a force from this field, again lifting them up off the surface. This could be the reason why Apollo astronauts in lunar orbit saw the horizon glow and dust fountains as they flew towards the sunrise. You can see this region for yourself – just look up at any time other than full moon to see the line between the lit and unlit regions. The dust only levitates to a height of 100 kilometres and there isn't enough for us to see it from the Earth. But it's there and it tells us that the Moon is moving through a cloud of electrically charged particles that are trapped inside our magnetic field and one aspect of our space weather.

The Moon isn't always inside the Earth's magnetic field though. Whether the Moon is inside or outside it depends on where the Moon is in its orbit. It's not that the Moon dramatically changes its distance from us – its orbit is very nearly

circular; instead it's the Earth's magnetic field that's highly distorted and the Moon moves in and out of it. From the solar wind
alone, the magnetic field is stretched out to a distance of as much
as 7 million kilometres on the nightside of the planet, well outside the Moon's approximately 380,000 kilometres' orbital
distance. On the dayside the magnetosphere is compressed to
70,000 kilometres above the Earth.

So, once a month, around the time of the new Moon, our
natural satellite pops outside the Earth's magnetic field for a few
days and finds itself sitting in the flow of the solar wind. The
next time you catch a glimpse of the new Moon, think about its
exposed place in the Solar System, being buffeted by the solar
wind. When the Apollo 17 lunar module touched down, the
Moon was just coming back inside the Earth's magnetic field.
The preceding Apollo missions kept inside the safety of the
magnetic field too.

With NASA, ESA and even some private companies planning
to send humans to the Moon again and also to Mars, understanding and predicting space weather is only going to get more
and more urgent in the decades to come.

14. What Comes Next?

The Sun has been around for about 4.6 billion years and has about that long again left on the clock. Our observations over the past few hundred years are just a blink in the Sun's lifetime. But the Sun is a changeable star and, of course, we want to know what the Sun is likely to do next, and how it will affect us.

EARTH FACING A MINI-ICE AGE 'WITHIN TEN YEARS' DUE TO RARE DROP IN SUNSPOT ACTIVITY

This was a UK newspaper headline that ran in 2010. I'm a solar physicist so it immediately caught my attention – partly because it talked about sunspot numbers, a subject very dear to me, but largely because of the sweeping claim that, because of the Sun, a mini-Ice Age was on the cards.

There does appear to be a background for this headline though. The idea that a drop in sunspot numbers could lead to a mini-Ice Age on Earth comes from an interval known as the 'Maunder minimum' – a time when sunspots were pretty much absent for around seventy years. It was Gustav Spörer, the German astronomer, who discovered this break in the normal pattern of the sunspot cycle. He died in 1895, though, before the solar physics community had accepted his proposal.

Spörer's ideas were taken seriously by one British astronomer, Edward Walter Maunder. He worked at the Royal Observatory in Greenwich, London, the leading institute for solar observations at that time, and he had a keen interest in sunspots. His analysis of the historical sunspot data led him to the same conclusion as Spörer. Together, their work established that during the years 1645–1715 there were virtually no sunspots on the Sun. History has forgotten about Spörer's contribution, though, and this seventy-year period when the Sun apparently turned off is named solely after Maunder.

The plot thickens though because the Maunder minimum is often said to have coincided with the Earth going through a period known as the 'Little Ice Age'. Evidence quoted to support this global cooling often includes the River Thames freezing over so deeply that frost fairs were held on them between 1600 and 1814, converting the river into an impromptu market to capitalize on the otherwise lost trade of the country's most economically important city.

Even William Herschel, who didn't know what sunspots were and didn't directly measure solar radiation across the Sun's cycle, pondered a possible link between temperatures here on Earth and the sunspot number. In 1801 Herschel turned to economics and saw that wheat prices were higher at times of low sunspot numbers. He published a paper where he reasoned that high wheat prices were driven by the scarcity of wheat, which in turn was driven by low terrestrial temperatures, resulting from a low number of sunspots. But are things really this simple?

Observations of tree growth using the thickness of tree rings does indicate a period of globally reduced average temperatures. However, this seems to have lasted for at least 500 years, ending around the mid-1800s. This is six times longer than the duration of the Maunder minimum. And even taking into account that sunspots were only recorded by European astronomers with telescopes from the early 1600s onwards, meaning that probably only three sunspot cycles were seen before the Maunder minimum began, over ten obvious solar cycles played out in the decades after the Maunder minimum and before the Little Ice Age ended – the Sun was active again for a long time before the Earth came out of its little freeze.

In fact, we know that the Sun was active before regular sunspot observations began and during the first half of this Little Ice Age because we can also use so-called 'cosmogenic isotopes' to study solar activity. These are generated by cosmic rays coming from our Galaxy – particles that we met earlier and that were seen as flashes of light in the eyes of some of the Apollo astronauts. And I mentioned that when they hit the Earth's atmosphere they produce nuclear reactions and it is these that generate the cosmogenic isotopes. They are then deposited in reservoirs such as tree trunks and ice sheets. And by drilling into those reservoirs we can measure their abundance in past times. The reason why this tells us about past solar activity is that the Sun's magnetic field helps protect the Earth from these cosmic rays, so the abundance of cosmogenic isotopes goes down when solar activity goes up, and vice versa.

A few alarm bells ring at this point. What I have just told you includes some broad-brush statements without any detail. I know that when I hear 'Little Ice Age' I think of global temperatures dropping, and it's easy to link stories of the Thames freezing over with a widespread drop in temperature. But we do need to be careful here – for a start, the tree ring measurements mostly tell us about the temperature in the growing seasons and almost nothing about temperatures in winter. And if one of the major rivers in a country froze over 200 years ago, and it doesn't today, is it fair to say that the temperature back then must surely have been significantly colder? Also, it seems that the cold spells during the Maunder minimum when the Thames froze were a feature of northern Europe only, and not a global event. And on top of that some of the hottest summers were recorded during that era. The temperature in London was slightly cooler during the winter of 1814 than it has been during recent winters. But there is more to the river freezing than just the temperature.

The main reason for the Thames's freezing over is that it was quite a different river back then. At the time of the frost fairs the river was wide, meaning that the water flowed slowly. Since the last frost fair, a significant amount of the river has been reclaimed to form the embankments. These embankments made the river substantially narrower, causing the water to flow much more quickly. The bridges across the river were different then too and they helped the river freeze. The old (medieval) London Bridge sat upon a series of narrow arches, making it a weir as well as a bridge, and this meant that any ice which did manage to form in the slowly moving flow could get caught, creating an ice dam that eventually led to the river freezing over upstream of the bridge. It was much easier for the Thames to freeze over a few hundred years ago than it is today. Things aren't quite as simple as the newspaper headline makes out. It's not as straightforward as: few sunspots = mini-Ice Age.

Behind the headline

But there is still some truth to this headline, even if the conclusion is incorrect.

We'll get to the 'rare drop in sunspot activity' in a moment. First we need to find out: can variations in the Sun's activity have a direct impact on the Earth's climate? If not an Ice Age, can changes on the Sun at least cause the Earth to globally heat up or cool down?

It's clear that the Sun is our major source of energy, so on a very simplistic level you might say that if the Sun becomes brighter we will have a warmer Earth and if the Sun dims the Earth will cool with it. We know that the majority of the Sun's radiation is coming from the photosphere and that sunspots are regions on the photosphere that emit less radiation than their sur-roundings because they contain cooler plasma. So the simplistic view would be to think that more sunspots mean a dimmer Sun and less energy coming our way.

Around the magnetic-field-intensive sunspot regions are smaller and weaker patches of magnetic field. These are the regions that form as the sunspot magnetic field disintegrates and disperses, and they are called 'faculae'. In the faculae, the upward convection of heat is inhibited, just as in sunspots, but their smaller size allows the radiation at their edges to keep the plasma in faculae hot. Sunspots are too large for this to happen. This combines with their relatively low plasma density, which enables us to see a little deeper into the Sun where temperatures are slightly higher. But since the walls of the faculae glow more brightly than the surrounding photosphere, faculae are brightest when they are seen close to the limb of the Sun.

The brightness of the faculae more than makes up for the dimming caused by the sunspots, resulting in a net increase in photospheric light being produced. At the maximum of the solar cycle the Sun is actually brighter than it is at minimum by about 0.1 per cent. At sunspot minimum the Sun is a tiny bit dimmer. So can a low level of sunspots affect our climate? And, if so, how?

It turns out that the number of sunspots can affect our climate. But not in the way we expect and not to a level that is significant. I hate to ruin the ending of the story, but it turns out that the Sun can cause the climate to change on the Earth, but to a much lesser extent than things like CO_2 emissions. The 0.1 per cent change in incoming solar radiation is a very small perturbation to the energy budget of the Earth's atmosphere compared to that caused by the increased trapping of heat by greenhouse gases. To put it plainly, the solar cycle's impact on the Earth's temperature is swamped by human-made climate change. If anything, despite the Sun's trying to cool the Earth with a diminishing number of sunspots and faculae, humans have still caused it to heat up. The effects of the

Sun are just one of many competing factors that influence the numerous complex processes which create our climate. And solar variability, at the moment, plays a very small part.

But to come back to the surprise. The first part is that the drop in photospheric light during solar minimum might be something of a red herring.

These warnings about how to interpret what happened in the past have been highlighted for several years now by Mike Lockwood, a professor of physics at the University of Reading. He revived an interest among the UK solar community in looking back through the long-term archive of data and not just focusing on what we have seen during our space age careers. And he encouraged us to put things into context: learning from the Maunder minimum but with an eye on regional temperature changes rather than global.

Mike looked to see whether there might be a link between the solar magnetic field and changes in the Earth's atmosphere that could lead to colder temperatures in northern European winters in the modern era. Using a whole range of space data he was able to see that changes in the strength of the magnetic field in the solar wind seemed to track winter temperatures in northern Europe. This got people thinking that the energy which flows out from the Sun might have a regional effect on our weather here on the Earth. As the saying goes, correlation doesn't imply causality, but it does mean the issue is worth looking into.

This work is interesting because it moves us away from the main point of the newspaper headline we saw at the start of this chapter, which just made us think about energy coming to us through the light of the photosphere. This is the most apparent vehicle for the energy, but there are other forms that are coming from the solar atmosphere. The solar wind is one way the Sun transports energy to us, and the gusty flows of this wind are varying all the time. But any variation is ultimately coming from changes in magnetic features in the solar atmosphere; after all, the solar wind is the expansion of the solar atmosphere.

So when Mike found an intriguing link between the magnetic field strength in the solar wind and northern European winter temperatures, it was an indication that we need to investigate whether changes in the solar *atmosphere* might be related to our colder winters. And as the magnetic field in the solar atmosphere varies, so too does the Sun's radiation across the wavelengths of the electromagnetic spectrum that our eyes cannot see.

This is an aspect that has been studied by Joanna Haigh at Imperial College London. Her work has focused on the ultraviolet part of the spectrum, where there is around a 10 per cent variation in the amount of light emitted between solar minimum and cycle maximum, with more ultraviolet light being emitted at solar maximum. The UV therefore has a fractional variation that is 100 times greater than the 0.1 per cent variation in visible light and heat from the photosphere. At X-ray wavelengths the variation across the cycle is even larger, being around 1000 times brighter at solar maximum than at solar minimum. Just as for the solar wind, this variation is driven by the evolving magnetic field in the corona. At solar maximum there are more regions of intense magnetic field emanating from sunspots, and the hot plasma trapped in these structures makes the Sun shine more brightly in ultraviolet and X-ray wavelengths.

We all know the photospheric heat and light penetrate to the surface of the Earth, and we all benefit from that. As we have seen right from the start of this book, most of the UV and the X-rays do not reach the surface, which is fortunate as they are harmful to humans. Most of the UV only reaches as far as the middle atmosphere (and in particular the stratosphere, about 20 kilometres up, where it is absorbed by ozone) and X-rays only reach the uppermost atmosphere (the so-called thermosphere, which is above 100 kilometres up).

Joanna's work considered the varying amount of ultraviolet light that the Sun emits and how this could have a regional effect at the Earth: from the minimum of the solar cycle to the maximum. Increased amounts of ultraviolet light falling on our atmosphere create more ozone, which in turn absorbs more ultraviolet radiation. But the important point is that when increasing amounts of ultraviolet light are absorbed by molecules in our atmosphere, the energy that the photons carry is also absorbed. So when more radiation is absorbed at these wavelengths the Earth's atmosphere in the stratosphere is heated.

The heating isn't uniform, though, and the equatorial stratosphere is heated more than that near the poles. This causes a gradient in both temperature and pressure and that drives a poleward motion of the gases, particularly in the hemisphere in which it is winter. And because the Earth is spinning, the poleward motion in turn causes an eastward wind because of the 'Coriolis force' - the 'jet stream' that straddles the boundary between the troposphere (the atmospheric layer that generates our weather) and the stratosphere above it. Meteorological science has shown that the jet stream has great implications for the weather in certain areas. When the jet stream moves north, warmer air from the south comes up to us, and when it moves south the opposite happens. The science is complex, but cutting-edge modelling is beginning to show how the behaviour of the thin air in the stratosphere can influence the lowest regions of atmosphere, and hence our weather.

By thinking about how invisible sunlight affects our atmosphere we might be able to rescue something from the newspaper headline. Joanna's work shows that a less active Sun could contribute to how many colder winters will be experienced by those of us living in northern Europe. And the models and the data both show that if that happens, other areas, like Greenland for example, would have a correspondingly higher fraction of warm winters. So far from causing a mini-Ice Age, a drop in solar activity could act to speed up the melting of the Greenland ice sheet! While the newspaper told us to think globally, we actually need to think regionally.

So that is it in a nutshell: the Sun transfers energy to the Earth in a range of different wavelengths, and a change in the magnetic activity on the Sun can have a knock-on effect which means that different regions on the Earth can be affected. But the impact of this is so localized that not only can it not cause an Ice Age, but also it cannot even compete with the human-made climate change that is dragging the temperature the other way.

So, that makes the headline wrong. A drop in sunspot numbers is not going to plunge the Earth into a mini-Ice Age. But what about the claim that a 'rare drop in sunspot activity' is happening?

Spot on

During the years around 2009 the Sun really *was* unusually quiet in terms of the number of sunspots that appeared. And this lull took most by surprise. Not Mike Lockwood and his collaborator Claus Fröhlich though. They had noted that the long-term solar activity levels had been going down since 1985 and that the trend was likely to continue. So what was happening?

As we saw before, the Sun's magnetic activity waxes and wanes over roughly an eleven-year cycle, which is marked by an increase and decrease in numbers of sunspots. Once the knowledge of a cycle was established, the data were backdated so that cycle 1 started in 1755. Solar cycle 23 had begun in 1996 and reached its maximum in the year 2000. Using the *average* cycle length of 11.1 years predicts that the Sun would be back in a minimum phase again by around 2007 – ending cycle 23 and starting solar cycle 24. Using the experience from the past cycles, scientists tried to predict what cycle 24 would have in store. When will the cycle 24 start and how large it will be?

To do this, the American National Oceanic and Atmospheric Administration, with support from NASA, created the 'Solar Cycle 24 Prediction Panel'. This panel gathered expertise from around the world and made use both of observations of the Sun and of models of solar activity. Several teams had to be brought together because a solar cycle prediction can be made in several different ways.

There are three ways to predict how a solar cycle is going to pan out, with increasing amounts of warning being traded off for accuracy in the predictions. If you wait for the cycle to actually start, the pace at which the sunspots appear can be an indicator of what will happen in the rest of the cycle, but by then the cycle is already under way. A step back involves looking at the conditions in the Sun before the cycle starts and seeing what information is there. Even further back, historical sunspot data, which span several cycles, can be used. But this method tries to tease out longer-term trends.

The most common prediction method is perhaps the most obvious and falls into the first of the three approaches: considering each solar cycle as an individual unit of the Sun's activity, with little interaction between cycles. Using this approach, previous sunspot cycles can be used to see whether there is a relationship between a cycle's sunspot number and the time and size of its maximum.

By studying lots of solar cycles it has been discovered that those cycles which show a rapid increase in the number of spots early on are likely to have large maxima, which arrive quickly. A sluggish start to the appearance of sunspots means a longer wait for maximum. This is known as the Waldmeier effect after the director of the Swiss Federal Observatory in Zurich who discovered it. The problem is that no prediction can be made until the spots of that cycle appear. After that we can of course fine-tune the prediction as more sunspots appear, but it would be helpful to have more warning about what a solar cycle is going to be like. For that, we have to look beyond just sunspots.

Another approach to solar cycle prediction falls into the second category and uses the Babcock–Leighton scenario, which looks at the magnetic field lines running between the poles of the Sun (the Sun's 'poloidal' magnetic field) because this is the seed magnetic field from which sunspots will later be produced. In theory, the poloidal magnetic field should be able to provide clues about the strength of the next cycle. Unfortunately, to cut a long story short, it is really difficult to actually track and measure the poloidal magnetic field. And whereas observations of sunspots go back 400 years, suitable observations of the magnetic field at the poles of the Sun really only go back to the 1970s.

Then there is the flow of the plasma itself, which is what drags the magnetic field around and amplifies it in the first place. The 'meridional flows' are the movement of plasma between the equator of the Sun and the poles. We can see these flows on the surface of the Sun transporting magnetic flux away from the equator, and we know that deep within the Sun the returning flows take the field back down again. And despite their slow nature, these flows represent a circulation of the solar plasma that appears to be involved in driving the solar cycle. It makes sense that because buoyant plasma then brings this magnetic field back up from the tachocline (after it has been strengthened by flows dragging the field around the Sun) to the surface to form sunspots, changes in the meridional flows providing this magnetic flux could have knock-on effects for the number of sunspots eventually emerging. But now there are so many factors and variables involved that the predictions become much more difficult.

But you don't just have to focus on the Sun. After all, we know that the consequences of its evolving magnetic field reach out into the Solar System. This has meant that one approach to predicting the size of a solar cycle, which has had some success, uses the level of geomagnetic activity, the disturbance to the Earth's magnetic field caused by the solar wind. This work has shown that there is a good correlation between the amplitude of a solar cycle and the level of geomagnetic activity at the *previous* solar minimum. That level of geomagnetic activity is set by the solar wind and the magnetic field embedded within it at the solar minimum and is thought to depend, to some degree, on the activity that took place in the preceding solar cycle.

This has given rise to a different approach to solar cycle prediction, which is to consider solar activity over timescales much longer than a single solar cycle. The long-term data are then taken as being continuous and periodic variations in the data are looked for. To take the weather around you on Earth as an example: every day it starts cold in the morning, gradually warms up during the day and then cools down again at night. The twenty-four-hour cycle of the temperature around you is a bit like the eleven-year sunspot cycle. If you look at more data you will see that not only does the temperature vary across a day, but that those cycles themselves gradually get warmer and cooler during the year (the seasons) and even those seasons can vary in the very long term.

In the sunspot numbers there is the eleven-year cycle, but superimposed on top of this cyclical variation are other periodic changes that occur over different timescales, such as the roughly eighty-year Gleissberg cycle (named after Wolfgang Gleissberg, who first suggested a periodic variation over this timescale – we'll come back to this later). If the collection of periodic changes can be picked out from the number of sunspots, they can be used to extrapolate into the future to give a probability of the size of future cycles before they have even begun.

Using this range of different techniques, the first long-range forecasts for cycle 24 were issued in 2006. A forecast at this time was difficult because the previous cycle hadn't yet ended, no cycle 24 spots had emerged to give an indication of the cycle progression. Despite the challenges, it was predicted that solar minimum would occur in March 2008 and the maximum of cycle 24 would be reached in 2011. The predictions forecast a cycle that would be moderate to large in size.

In 2009, a ripple of excitement started to flow through the solar physics community. The first spots of cycle 24 had been seen in January 2008. We knew they were from cycle 24 because they were of the reverse magnetic orientation (remember, it flips every cycle). Sunspots of cycle 23 that appeared in the northern hemisphere had a positive magnetic polarity in the leading spot, whereas those of cycle 24 are negative. The sunspots were given the name '10981' (sunspot pairs are numbered in a sequential way), an inconsequential name but one that now represents a change of season on the Sun.

Within three days the spots were gone though, as the magnetic field was quickly dispersed into the surrounding photosphere. Even though the sunspots were small and only lived for a few days, they were important and signalled that cycle 24 was here. For a while the spots of cycle 23 and 24 overlapped. It's completely normal for the Sun to do this. Eventually the spots of cycle 24 began to dominate over the spots of cycle 23 in September 2008.

Further cycle 24 spots, however, were very slow to appear and this is why the solar community grew excited. The Sun seemed to be quieter than it had been for the previous 100 years. Things were exceedingly quiet. With very few spots there was no significant magnetic activity on the Sun, and the solar wind dropped too. Its pressure fell by 22 per cent and less magnetic flux was being carried away from the Sun. The weak magnetic conditions on the Sun propagated into the Solar System.

This reduction in the dynamic pressure of the solar wind could have an effect all the way out to the edge of the Solar System. The bubble of the heliosphere that swells out well beyond the planets is shaped and controlled by the forces of the solar wind and changes in the wind ripple outwards so that the heliosphere is perhaps smaller now than it has been in past cycles. This may very well have had the consequence that the Voyager 1 spacecraft on its mission to reach interstellar space didn't have so far to go!

The low levels of solar activity forced us to realize that the abundant sunspots and the wonderful explosive activity that we had watched and been fascinated by during the space age might not be the defining feature of our Sun. Some even speculated that the Sun might be headed for another Maunder minimum phase.

Even though sunspot numbers were low, by 2009 enough had been seen for them to refine the cycle prediction methods. The legion of different approaches were amalgamated once more, and a revised prediction was issued by the Solar Cycle Prediction panel in May 2009. This time they suggested that solar maximum would occur in May 2013 but that cycle 24 would be smaller than average.

As 2013 rolled along the sunspot numbers continued to disappoint, prompting a new speculation about solar maximum based on previous cycles that looked similar to cycle 24 during the rise phase. It seemed that sunspot maximum would now be delayed until late 2013, 2014 or maybe even 2015. But it turned out not to be quite that bad. Solar maximum occurred in late 2013 and continued into 2014. The hemispheres peaked at different times, which is not unusual. The rate at which sunspots appear in different hemispheres doesn't have to be the same and this means that one hemisphere will reach solar maximum before the other.

We now know that the Sun has been at its quietest since 1906. You could indeed say that it was a 'rare' drop in sunspot numbers, certainly on the scale of a human lifespan or the duration of the space age. Score one for the headline!

But it may not be rare in the lifespan of the Sun. For me, it was fascinating to watch the Sun show a different side from the one I am familiar with, but the recent activity level goes to show that we shouldn't judge the Sun on what we see during our relatively short careers. Why should such a short snapshot be a representative view of a star that has been living out its life for 4.6 billion years already? As cycle 23 waned the Sun entered a quiet phase that was deemed unusual by solar physicists who had been studying the Sun from the twentieth-century perspective. What can we learn about predicting the future from looking at the long-term activity?

The long view

Sami Solanki, a solar physicist and director of the Max Planck Institute for Solar System Science in Germany, introduced me to looking at the long-term solar activity: going further back in time than just the sunspot observations that were carried out in Europe. He was able to do this because there are ways to investigate the Sun's long-term magnetic activity by using proxies rather than direct measurements. We met this before. The pulsing in strength and complexity of the Sun's magnetic field is felt at the Earth by the level of shielding we get from galactic cosmic rays – high-energy charged particles.

It's a bit like how the Earth's magnetic field protects us from the Sun's harmful particles: the Sun's magnetic field protects the entire Solar System from particles coming in from other stars. We are located inside the Sun's atmosphere, and how many galactic cosmic rays can penetrate into the Solar System and reach us on the surface of the Earth is affected by the strength of the Sun's magnetic field. Fewer are able reach us at times of solar maximum, when the Sun's magnetic field is at its strongest.

As we also saw in the previous chapter: high-energy cosmic ray particles leave their calling card through the production of the radioactive isotopes carbon-14 and beryllium-10. Radioactive particles are unstable and transform into other particles over time by breaking up. The half-life of carbon-14 is 5730 years, meaning that in that amount of time half of the particles in any collection will have broken up, and the half-life of beryllium-10 is 1.5 million years. Since these isotopes are created by cosmic ray impacts with oxygen and nitrogen in the stratosphere, and are ultimately stored in tree rings and laid down in layers of snow, they act as a time capsule through which we can study the past.

Sami used these pieces of data to reconstruct the level of solar activity back over the past 11,400 years. Despite the uncertainties in the reconstruction of the solar activity from such data and the difficulties in trying to pick out an eleven-year cycle, long-term variations in the levels of these radioactive isotopes provide a way to look at the Sun's activity over hundreds or even thousands of years, which in turn can be used to help understand what activity levels we might expect from the Sun in the future.

These datasets show that the Sun went through other Maunder minimum phases before the European sunspot records began. This includes the Wolf and Spörer minima. The new perspective that this long-term view gives us shows that the Sun has grand minima which occur irregularly across the millennia. Over the last 9000 years of the history of the Sun sixty-six occasions of relatively high solar activity can be picked out of the long-term modulations in cycle size. They are somewhat arbitrarily defined but they are taken to represent times when the eleven-year solar cycles are large for several cycles in succession. These times of heightened activity have come to be called grand maxima.

It seems that whilst judging the Sun on its eleven-year sunspot cycle is comfortable for human lifetimes, it is rather short-sighted on solar lifetimes. The complex ebb and flow of the sunspot numbers and the magnetic Sun have been taken for granted. But there are far longer-term trends that we are only beginning to appreciate. It seems that the state of the plasma flows and magnetic field movements evolve over timescales well beyond what we will ever be able to witness directly. We are probably only now just starting to gather the right information through our telescopes in space and on the ground. And we need to keep gathering these kinds of data for many generations to come. But with our current data we can still try to understand the solar rhythms that have an impact on our life here on Earth.

Scientists are still trying to work out how each eleven-year sunspot cycle can affect the next one, particularly when the cycles become almost non-existent. It is fascinating that the solar cycle can apparently switch off for several decades so that no strong magnetic fields appear at the surface to form sunspots. But then, seemingly from nowhere, the Sun recovers and the cycle re-establishes itself. The cycle must have been hidden the whole time, somewhere. This is what appeared to happen during the Maunder minimum. It's a puzzle. What can we confidently say about the future?

How odd is our current Sun?

Well, Sami Solanki studied the long-term magnetic activity and found that although cycle 24 is weak, the levels of solar activity which we have witnessed over the last seventy years are exceptional. We now know that the last time the Sun was at such a high level of activity appears to have been over 2000 years ago. We might be trying to understand the Sun at a time when it is very much atypical.

Sami suggests that during the last 11,400 years the Sun has only spent 10 per cent of the time being in such an active state. It seems we have been privy to a rare event. Our observations of the Sun

during the space age, the time when solar physics rapidly progressed, seem to have coincided with the Sun being in a
grand-maximum state. But now we appear to be coming out of
the grand maximum of solar activity. This decline in overall solar
activity appears to have been happening since around 1985.

And, making a prediction from the cosmogenic isotopes, Mike
Lockwood suggests that there is roughly a 10 per cent chance of
the Sun going into a Maunder minimum in the next forty years
and a 45 per cent chance within 150 years. If solar activity declines
over the coming cycles then we may be looking at a dimmer Sun
in the decades to come, but this is expected to have only a very
small impact on global temperature. If the Sun goes into a grand
minimum we could expect to see a cooling of 0.1–0.3 degrees
Celsius as compared to the warming by 3 degrees expected from
the rise in global temperatures that we are currently experiencing.
But we may also see a larger fraction of relatively cold winters in
Europe because of the effects of UV light on the stratosphere.

The lessons learnt from making predictions about solar cycle
24 and seeing how things played out in reality tell us a great deal
about what we do and don't yet know about our local star. It's
been a thrilling time to take a step back and reflect on the bigger
picture; to focus not just on what the Sun is doing now but also
on what it might do in the years, decades and centuries ahead.
There is another way to expand our horizons though, and that is
to compare the Sun to the other hundred billion or so stars in our
Galaxy, the Milky Way. This is the Sun's extended family. What
can we learn from considering our Sun as a star among so many
billions of others?

Conclusion: Our Special Sun

As our local star, the Sun holds a special place for those of us on Earth. Not just because it makes life here possible in the first place, but also because its proximity means it absolutely dominates the sky: literally outshining all other stars. It is easy to become obsessed with our local star and forget that it is part of a much wider nuclear family.

But how does our local star fare on the galactic scene? Sure, it rules over our Solar System, but how does it compare to the hundreds of billions of other stars in our Galaxy? Scientifically, we can use our knowledge of the Sun to help us understand other stars, and indeed surveying so many other stars can give us insight into, and context for, the Sun. Personally, I'm quite keen to find some validation for my belief that my favourite star is still special among all the others in the Universe.

I first realized I had an innate want to champion our local star (in the same way you would support a local football team) when I read about star R136a1. Admittedly, it hasn't got a very catchy name. With billions of stars out there, astronomers don't try to give them all individual names like Betelgeuse or Alpha Centuri, so new discoveries are given a catalogue number. The letters tell you which catalogue a star is in (based on which observatory or project discovered or logged it; the R in R136a1 tells us it was discovered by the Radcliffe Observatory in Pretoria, South Africa) and the number is its position in that catalogue. Anything after that number, in this case 'a', tells you that when astronomers looked again later, using better telescopes and techniques, at what they assumed was one star, they realized it was actually a few stars huddled together, which are then given the suffix a, b, c, etc. And when that happens again, R136a is subdivided and the stars are labelled R136a1, etc.

R136a1 lives in a cluster of stars about 165,000 light years from Earth. For a start it is much hotter than our Sun: whereas the Sun's photosphere is around 6000 Kelvin, R136a1 is about 50,000 Kelvin. And it is much younger, with an age of only 1.5 million years compared to the Sun's 4.6 billion (the Sun is 3000 times older, in other words). But what really caught my attention is how massive R136a1 is: it has a mass 265 times greater than that of our Sun. It helps to look at other stars when trying to understand our own.

The main sequence

Before we look more closely at R136a1 though, we need to take a moment to orientate ourselves. When it comes to looking at the range of stars in the Galaxy it's useful to refer to stars that are collectively known as 'main sequence' stars. These are the stars that Annie Jump Cannon classified at that incredible rate of three a minute. Broadly speaking, main-sequence stars can be thought of as stars that are all being powered the same way: they are stars in whose interiors hydrogen is fusing into helium.

As you will recall from chapter 3, different types of star within the main sequence are labelled with the letters O, B, A, F, G, K and M in that odd order. O-type stars are the hottest and M-type the coolest, so R136a1's blistering-hot surface makes it an extreme example of an O-type star. The Sun's roughly 6000 Kelvin surface makes it a G-type star, not far off the middle of the OBAFGKM sequence: so in terms of temperature the Sun looks pretty average.

It's not really fair to compare the mass of the Sun to the mass of R136a1 though. R136a1 isn't your normal star. After it had first been spotted, a study led by Paul Crowther, Professor of Astrophysics at the University of Sheffield, worked out that as well as currently being 265 times more massive than the Sun, it would have started out with much more material than that. Since it formed over 1.5 million years ago, strong winds from the star's atmosphere have been continually eroding its outer layer. At birth R136a1 is thought to have been 320 times more massive than the Sun and this makes it a record-breaker: it's currently the most massive star that we know of in the entire Universe. But what really makes it remarkable is that it is twice as massive as was previously thought possible for a star.

The reason why there should be an upper limit to a star's mass comes down to how stars form: from the accumulation of material in a shrinking cloud of gas and dust called a nebula – the same basic process which produced our Sun and that we visited in chapter 3. And logically it would follow that if more material falls in during this process, the growing star will become more massive.

But nebulae are finite in size, and when the supply of material stops, the star ceases to grow any further – you can't keep building a house if you run out of bricks. But construction can stop for other reasons before you run out of raw materials, and likewise there's another mechanism that can prevent a star endlessly accumulating mass, and which derives from the star itself and how it interacts with the surrounding nebula: at some point the star starts to push back against the material that gravity is pulling in and so there is an upper limit to how massive a star can be.

During the formation of the Sun, the pressure of the high-temperature plasma works against gravity. Since the two became balanced, the Sun has kept at a constant size. But there is another way to push back on the material, and that's by using light. Our Sun isn't creating enough radiation to have a significant effect so only a small outward pressure comes from the photons that are working their way out and into the Solar System. But the most massive stars in the main sequence are also the most luminous. They do produce enough light to exert a pressure on the surrounding material of the nebula, so that it starts to push the cloud back.

It was thought that this effect becomes important for stars that are much more massive than our Sun, perhaps around 150 times as massive, so this would set the upper limit on the mass. Or at least that was what was thought until R136a1 was discovered: it extended the range of masses that we know stars to have. Scientists are now re-evaluating their theories to try and find a more accurate understanding of how stars form and interact with the nebula that birthed them, one that fits the stars we see. This is how science advances: theories are revised in the light of new evidence. And when I spoke to Paul Crowther about R136a1 he told me that, although it is still a record-holder, scientists have since found another ten stars in the R136 cluster that have around 100 times the mass of the Sun.

With this in mind, let's put aside extreme examples like R136a1 and look to see whether the Sun's mass is special or not. We know it is certainly not a high-mass star, but how does it compare to the lowest-mass stars? The minimum mass a star can have is determined by the conditions needed to switch a star on. And it's been known for some time that there is a lower limit to the mass that a star can have.

Stars need a sufficiently high temperature and pressure in the core for fusion to occur. And to get to a high temperature and pressure requires enough mass to have been pulled in under gravity. Taking this into account means that the smallest stars will have around 8 per cent of the mass of the Sun. Any less than this and the temperature and pressure will not be sufficient for hydrogen fusion to take place. These small stars are called 'red dwarfs'. They are less massive and cooler than our Sun – they have a surface temperature in the range 2500–4000 Kelvin – giving them a red glow.

So we have giants like R136a1 at the high-mass end that are over 260 times more massive than our Sun, and dwarfs at the low-mass end which are twelve times less massive than the Sun. From this I would say that when compared against all main-sequence stars the Sun is a very light flyweight – it's puny and not impressive at all.

But that's just the range of star masses. I've been very careful so far to say 'massive' and not large. There is a big difference between how much mass a star has and how much space it actually takes up. And it's perhaps easier to conceptualize and compare stars in terms of their physical size. How does the Sun rate in this measure?

If the sequence of stars were placed side by side in a line-up, ordered by their size, the smallest red dwarfs would be around 10 per cent of the width of the Sun; R136a1 on the other hand, despite its vast mass, would be only about thirty times wider: if R136a1 replaced the Sun it would extend less than halfway to Mercury. So, in a size line-up, the Sun looks neither particularly small nor large – it seems very average.

Size is actually a slight distraction: it's the mass that really matters because it sets the age that the star will reach. Our Sun is currently halfway through its nine-billion-year lifetime. The lowest-mass stars, red dwarfs, are the coolest and use up their hydrogen supply very gradually: fusion runs slowly in their interiors. So red dwarfs aren't terribly bright, but they do live out long lives. So long, in fact, that the lowest-mass red dwarfs could have a lifetime that is longer than the current age of the Universe – the Universe is 13.8 billion years old. Stars more massive than our Sun have higher central temperatures and pressures, which means they fuse hydrogen more rapidly, shine more brightly and run out of their supplies much more quickly. An O-type star that is forty times the mass of the Sun typically lives for only 5 million years. They burn bright and die young. In this sense the average nature of the Sun in terms of its lifetime makes it a Goldilocks star. It has just the right amount of mass.

But what has always made the Sun special to me and the generations before me, since the era of Hale and the Mount Wilson Observatory, is its magnetic field. This is what turns an otherwise bland ball of plasma into a dynamic and active star. Our Sun starts to look much more interesting when we know that its magnetic field powers solar flares and coronal mass ejections. And that the magnetic field stretches across the Solar System, extended out through the gusty flow of the solar wind. From space weather on Earth to the edge of the heliosphere, the Sun exerts its presence because of its magnetic field. Do other stars have magnetic fields and do they have magnetic activity too?

Stellar magnetism

For me the discovery of the Sun's magnetic field by George Ellery Hale in 1908 marked the birth of solar physics and, but for scientists of that era, it would have made the Sun unique, because it became the first star known to have a magnetic field. It was indeed the first time any magnetic field was detected beyond the Earth. But the Sun is fundamentally no different to other stars in the main sequence – they are all spheres of plasma fusing hydrogen nuclei to helium nuclei in their cores – which raises the question: do other stars have magnetic fields too? It turns out that answering this question is a very challenging task.

The Sun has a relatively easy magnetic field to detect because of its proximity, which allows us to measure the photospheric features like sunspots. Other stars are just dots of light; there are exceptions – like Betelgeuse, which is so big and relatively close in astronomical terms that its size and shape can be made out with a very powerful telescope. But in most cases stars are points of light and this makes it incredibly hard to find out whether they harbour a magnetic field.

As Hale and others realized, magnetic fields reveal themselves in a star's light. A north magnetic pole imprints a certain signature in the spectral lines and the south pole imprints a different one. On the Sun the north and south poles can be resolved, but for other stars the light from these different magnetic field patches is merged together – and the imprint of the magnetic field can be lost. There are other things that can affect the spectral lines too, such as the rotation of the star and the effect of the temperature of the emitting plasma. With so many unknowns it is hard to pull out only the information about the magnetic field. So it was a major triumph when, in 1980, a magnetic field was detected on another star. It had been widely assumed that some stars would have magnetic fields, but detecting such a field made this assumption a scientific certainty. And the observations of stellar magnetic fields that followed means today magnetic fields on other stars are seen as commonplace and are fundamental to studies of them. Our Sun is not a special case.

Blood, sweat and flares

Perhaps our Sun stands out because of the activity that the magnetic fields create? The use of energy stored in the magnetic field produces the most magnificent flares and coronal mass ejections. And this storage and release is all part of the solar cycle. Curiously, it was possible to detect a *cycle* on other stars before a stellar magnetic field was first detected. This was because there are proxies that can be used to infer that a star has a magnetic field (without detecting the magnetic field directly) and whether it varies. This technique uses light from a star's chromosphere, the amount of which changes if the star's magnetic field changes. According to data gathered at the Mount Wilson Observatory, some stars showed no regularity in their inferred magnetic activity, but about 60 per cent of the stars observed did behave in a cyclic way which indicated they had a magnetic field and that it was cycling too. Meaning that our Sun isn't alone in having a magnetic cycle.

Given that there are magnetic cycles on other stars, do their magnetic fields power stellar eruptions and flares? The magnetic cycles of other stars are peppered with bursts of activity that look very much like solar flares. We observe flares on the Sun, using wavelengths across the electromagnetic spectrum; creating images that can see the details of the flare, we can infer the changing configuration of the magnetic field, and we monitor the total output of light to make a so-called light curve that allows us to track the intensity of the radiation. Other stars cannot be spatially resolved, but the amount of radiation they emit across the electromagnetic spectrum can be measured and used to produce light curves for them too. When these light curves are studied I am sorry to report that they show the same transient increases in brightness that the Sun shows when a solar flare occurs. Our Sun is not unique in this way either.

But there are ways in which our Sun might be more interesting. Data collected by a satellite that was actually designed and launched to study planets outside our Solar System suggest that we may not have seen all that the Sun can do.

NASA's Kepler satellite was launched in 2009 to keep an eye on 100,000 stars in the constellations of Cygnus and Lyra. That vast number provides a big pool to dip into when looking for the almost imperceptible dips in light as planets pass in front of a star. But in this pool are many stars different in age to our Sun but similar in terms of their surface temperature and rotation rate. They provide a way of investigating what our Sun might have looked like in the past or might look like in the future. And looking at the light emitted by these Sun-like stars shows that they have flares just like our Sun does. While this was known about before Kepler, Kepler provided a larger number of stars to study and it showed that we might not yet have witnessed the full potential of what the Sun can do.

The Kepler data show that flares on these Sun-like stars can be 10–1000 times more energetic than those seen on our Sun. The activity on our Sun seems dramatic to us, but is dwarfed by these so-called super-flares. But such large flares are thought to be rare, happening perhaps once in 1000–5000 years. I am glad that our Sun doesn't regularly produce super-flares. This is probably the reason why life was able to evolve. If this makes it boring, then good! There is no point having an exciting local star if you can't evolve to enjoy it.

Amongst the main-sequence stars, all types produce flares, from the most massive stars to the least massive. Flares have even been observed on very young stars that haven't yet started fusing hydrogen nuclei in their cores and made it to the main-sequence phase. It also doesn't matter if the star is one of a multitude of stars or even in a binary system of two stars orbiting each other – they produce flares. Our Sun is not unique when it comes to having a magnetic field, a magnetic cycle or magnetic activity like flaring.

End of the Sun

Maybe the Sun will be special when it dies?

The word 'supernova' is widely known to describe the explosive and near-catastrophic end to a star's life. But whether or not this happens depends on how massive a star is.

When most high-mass stars come to the end of their lives they first become a red supergiant star. Betelgeuse – in the constellation of Orion and a star that I've mentioned before – is the best-known example of a red supergiant. It's moved beyond the hydrogen-burning phase of its life and is currently fusing helium into carbon in its core. It has over ten times the mass of the Sun but it has swollen in size to become around 1000 times bigger as well. If you replaced the Sun with Betelgeuse it would engulf all the planets as far out as Jupiter.

Stars go through different phases of fusion for reasons of supply and temperature. When an element that has been fusing within a star is completely used up, the star contracts because it is no longer exerting sufficient thermal pressure outwards to counteract gravity. But the collapse raises the internal temperature and eventually a point can be reached where a different element will start to fuse. The larger the star, the higher the temperature that can be reached, and this means that a succession of elements are used in a sequence of fusion phases.

The insides of these massive stars near the ends of their lives resemble a giant stellar onion with the heavier elements sinking towards the core. But there is a limit. Fusion can occur as far up the periodic table as the element iron; for iron and beyond, energy needs to be absorbed, rather than emitted, in order for particles to be fused together. Therefore the most massive stars can end up having iron cores. I can't help thinking back to those early studies of the composition of the Sun previously described – even Eddington had thought the Sun was mostly made of iron. Our Sun isn't massive enough to forge this element, but there are stars out there that do.

Once an iron core has been created the fusion stops. This is the end of the line for the star. And with no fusion to release energy to counteract gravity, the core collapses. There is an unimaginable release of energy and a rebound that blasts the layers of plasma surrounding the core out at thousands of kilo-metres per second, leaving behind a remnant known as a 'neutron star'. The name gives away what this star is made of — neutrons, created when the protons and electrons in the collapsing core were squeezed together. These types of star are around the size of a city and so dense that just one teaspoon of their material would weigh a billion tonnes. And the collapse leaves the neutron star spinning at a phenomenal rate — perhaps thirty times a second!

In the case of even more massive stars the neutron star itself can collapse yet further and turn into a black hole: a region of space where gravity is so strong that not even light can escape. And R136a1? Well, this star is so massive that it will live a short life and have plenty of hydrogen and helium when it dies. Rather than fusion proceeding in a controlled way until the stocks are depleted, R136a1 is likely to enter a phase when runaway nuclear fusion creates a thermonuclear explosion that cannot be contained and it will completely blow itself up. Nothing will be left behind.

And our Sun?

Once it enters the phase of helium burning, our Sun will swell to become a red giant. The Sun as we have known it will be gone. And so, too, will we. The Sun will probably inflate in size to the point where it has expanded beyond the orbit of the Earth. And it will start to shed its outer layers in fits and starts and form a planetary nebula. With fusion finished, the core that is left behind collapses, unable to fight against gravity, and shrinks down to become an object that is the size of the Earth but has a temperature of around 100,000 Kelvin — our Sun will then be known as a 'white dwarf'. Without a power source the white dwarf simply cools down over the course of the next billion years or so, fading from sight. A rather graceful ending.

One of a kind

I still believe that our Sun is special: for what it is in its own right but also because it created an environment that allowed life to form and thrive. As far as we know, the Earth is the only place in the Galaxy, and maybe the Universe, where life exists. And as of 2015, even though water has been detected on Mars, no extant or extinct life has been found there. Meanwhile, on Earth, ancient rocks and microfossils suggest that life emerged after only a billion years, and the Sun's magnetic field may well have helped this happen.

The solar magnetic field extends out into the Solar System, creating an invisible shield around us which deflects high-energy cosmic rays coming from our Galaxy. We met these particles earlier because their tremendous speeds and energies mean that if they reach us they can penetrate the human body. When this happened to some of the Apollo astronauts, the particles revealed themselves by causing flashes of light in their eyes. But if the astronauts hadn't been within the Sun's magnetic envelope they would have seen an awful lot more flashes. The Sun's magnetic field provides a protective layer, without which the bombardment of galactic cosmic rays could have damaged early living cells and torn DNA apart. The consequences could have included cell mutations – taking evolution down a different path.

The reason why you are able to read this book is because life did manage to thrive and eventually become intelligent life. We took around 3.8 billion years of evolution to arrive and have spent just the last few hundred years studying the science of the Sun. Perhaps the existence of human intelligence makes the Sun special.

Using just the one example we have, we could conclude that stars need to live for billions of years for intelligent life to form.

So extrapolating wildly would mean that the most massive stars, that only live for millions or tens of millions of years, will not last long enough for intelligent life to emerge on any planets that have managed to form around them. And that's ignoring the impact that their searing surface temperatures and harsh radiation might have on any life that does manage to develop.

So let's only consider stars that live for over a billion years, i.e. stars slightly more massive than our Sun down to red dwarfs. The formation of life anywhere in the Universe is thought to need water. Liquid water is such a good solvent for chemicals that it is thought to be where chemistry turned into biology and life began. And for water to exist in liquid form needs the right conditions for it to have the right temperature. We know from experience that being on a planet orbiting a Sun-like star can provide these conditions. Can these conditions exist elsewhere?

The radiation from any star spreads out in all directions, so the amount of energy that a planet receives depends on how bright the star is and how far the planet is from it. Doing this calculation for the Earth predicts that our planet should have an average temperature of 7 degrees Celsius – not bad in terms of what we see, and definitely the right temperature for liquid water. In reality the Earth's temperature is warmer than our simple calculation predicts because we have an atmosphere that traps heat from the Sun. And the different natural environments around the globe and variations in the amount of energy that falls on particular parts of the planet create a temperature that varies from place to place. But averaging out across these diverse features gives a temperature of around 16 degrees Celsius.

A planet in orbit around a cooler red dwarf would have to be much, much closer in to its parent star than we are to be warm enough to have liquid water – perhaps as close in as the planet Mercury is in our Solar System (about one third of the distance between the Sun and the Earth). So even though it looks unlikely that a planet in an Earth-like orbit around a red dwarf would be warm enough for liquid water to exist, the presence of an atmosphere would be a help. An atmosphere with greenhouse gases, such as carbon dioxide or methane, would trap heat and raise the planet's temperature. Possibly to the point where any water ice melts.

Only child

The modern view of star formation is that when vast gas clouds give birth to stars it isn't on a one-to-one basis. They spawn several, perhaps hundreds, of new stars. Perhaps our Sun is unique in being an only child? To be able to answer this question needs more than a stellar line-up. A solar sibling doesn't need to have the same mass, size or surface temperature as the Sun. But it does need to be made of the same material; this is the forensic evidence needed to show that two stars share a common origin. The cosmic DNA is found in the stellar spectra, which show the chemical composition of the star – and after a study of nearby stars a solar sibling eventually turned up!

The Sun's sibling was found because it has the same composition as the Sun but also because it is moving along a path that makes it look like it came from the same part of space. The star is known as HD162826 (HD standing for the Henry Draper catalogue) and it's about 110 light years from the Sun and 15 per cent more massive. This is an exciting find. HD162826 has had plenty of time to wander off into the Galaxy and potentially be lost from our view. It's a fairly bright star that remained fairly close by and so, if you have a good telescope, you may be able to find it too, in the constellation of Hercules.

But even after the discovery that the Sun is part of a nuclear family, it's still my favourite.

As simple as a star

It is an incredible time to be alive and working in solar physics. The Sun is going from a period of unusually high activity to one that may be unusually quiet – and right at a time when we have the technology both on Earth and orbiting through space to watch it all happen. Every morning when I get to work I know that there are literally terabytes of new data about the Sun that I can look through. And with Voyager 1 at the edge of the heliosphere moving out and Solar Orbiter and Solar Probe Plus about to be launched to *see* the Sun from close up, we're continuing to widen our view of this amazing star. We are only going to have more and more solar revelations in the years to come.

Before Eddington focused our attention on understanding something as simple as a star, the discoveries about the Sun were coming from all directions from every field of science. There was no such thing as solar physics – there was just scientific thinking applied to the Sun. And today that is still true: when I look at the range of scientists and engineers I call my colleagues, I realize that there is no such thing as a 'solar physicist' (despite what my business cards say). Our community contains all sorts of experts who happen to be interested in the same celestial body.

And the work carries on. We still have fundamental questions about the physical processes that govern the solar plasma and magnetic field. We are desperate to have telescopes that can see more detail, so that the tiny regions where magnetic reconnection occurs might eventually come into view. Magnetic reconnection is a fundamental process in the solar atmosphere but so far we have never been able to see the plasma involved directly. And at the opposite end of the size-scale there is much to learn about the solar dynamo, crucial in understanding the heartbeat of the Sun and knowing what it will do in the future.

So despite the Sun being the most studied and understood of all stars, a detailed physical description of many processes remains to be developed. What we ultimately hope for is a grand unified picture: a systems approach to understanding the Sun in which no one part of the Sun is considered in isolation. And to achieve this we need observations that give a seamless coverage throughout the atmospheric layers – something that hasn't yet been achieved but might become possible with future generations of satellites and telescopes.

If Eddington were alive today I think there could be no question but that he would be impressed by just how far we have come, and perhaps excited that a whole new set of questions has been raised for future generations to work on. We aren't yet at the point where we fully understand something as simple as a star. But we have learnt an awful lot in trying.

Appendix: How to Safely Observe the Sun

There is a wonderfully simple way to view the Sun which is safe and doesn't involve anything other than a small hole in a piece of card – no mirrors or lenses are needed. The device is called a pinhole camera and to make one get a couple of pieces of card and some aluminium foil. Cut a hole about 2 centimetres square out of the centre of your sheet of card. Place the aluminium foil over this opening and tape it down at the edges. Using a pin, make a small hole (1 millimetre) in the centre of the foil. Hold up the pinhole to a bright object with the second piece of card (which acts as the screen) behind it and you have a pinhole camera! You may want to test the camera first because getting the distance between the pinhole and the screen is a bit tricky. To test the camera, hold it a few centimetres away from a lit candle with the piece of white paper on the other side of the card to the candle. You should see an image of the flame projected onto the piece of paper. The image of the flame will be upside down.

Once you have mastered the use of a pinhole camera you are ready to use it to view the Sun. Outside on a cloud-free day place the screen on the floor and hold the pinhole above it. Move the piece of card with the pinhole up and down over the screen in the direction of the Sun to bring the image into focus. To get a projected image of the Sun that is about 1 centimetre in diameter, the other sheet of paper will need to be about 1 metre away. But experiment. You'll see the disc of the Sun projected through the pinhole and onto the screen.

Using a telescope to observe the Sun is much more risky. The aim of a telescope is to magnify the image of an object, but it also gathers more photons than the eye can by itself because it has a larger opening than our eyes. More photons equal more eye damage. So when it comes to studying the Sun, you'll need either to let the light pass through the telescope so that it falls onto a flat surface and creates a projected image of the Sun, or to use a purpose-made filter from a specialist supplier. My advice is that, before purchasing this equipment yourself, you should get some support and information from your local astronomical society. They will have telescopes to view the Sun and can show you how to do this safely. To find your local society visit http://fedastro.org.uk/fas. Or you can join the Society for Popular Astronomy and get support all year long: http://www.popastro.com.

Glossary

convection zone
: Region in which energy is transferred mostly by the convection motions of the plasma located between the radiation zone and the photosphere

corona
: Outermost atmosphere of the Sun seen in visible light during a total solar eclipse and in X-ray and extreme ultraviolet images from space

coronal mass ejection
: A sudden eruption of magnetic field and plasma from the Sun's atmosphere into the Solar System

electron
: A negatively charged subatomic particle

flux rope
: A bundle of twisted magnetic field lines and electric currents

helioseismology
: The study of the solar interior using sound waves observed in the photosphere

heliosphere
: The magnetic and plasma bubble created by the outflowing solar wind and encompassing the Solar System

ion
: Positively charged particle formed from an atom that isn't electrically balanced

light year
: The distance that light travels during one year – 9.46 million million kilometres. Often used as a unit of distance in astronomy

magnetic helicity
: A quantity that describes how twisted, linked and distorted a magnetic field is

magnetosphere	The region of space surrounding an astronomical body which is filled by the object's magnetic field
nuclear fusion	A nuclear reaction in which light atomic nuclei fuse together to form a heavier nucleus, releasing energy in the process
nucleus	The central part of an atom or ion that contains protons and neutrons
photosphere	The 'visible' surface of the Sun and the deepest layer that we can see in visible light
plasma	A gas consisting of separated electrically charged particles (electrons and ions) in roughly equal propotions so that there is no overall charge
poloidal	A magnetic field which has field lines that are parallel to lines of longitude
radiation zone	Region in which energy is transferred mostly by photons located between the core and the convection zone
refraction	Change in direction of a wave caused by different parts of the front of the wave moving at different speeds
solar dynamo	A collection of processes that produce a varying magnetic field in a star. A dynamo involves the conversion of kinetic energy in the gas motions into magnetic energy
solar flare	A sudden burst of electromagnetic radiation in the atmosphere of the Sun. The glow can last from minutes to hours and has a strong X-ray component
solar limb	Edge of the disc of the Sun
solar nebula	The cloud of gas and dust from which the Sun and the other Solar System bodies formed

spectroheliograph	An instrument that creates an image of the Sun in one wavelength only
spectrum	The entire range of wavelengths of light emitted or absorbed by a substance
sunspot	Relatively dark and cool region in the photosphere that is created by an intense magnetic field
tachocline	Region between the rigidly rotating radiation zone and the convection zone where rotation changes with latitude and depth
toroidal	Having the shape of a torus

Bibliography

Articles

Abreu, J. A., Beer, J., Steinhilber, F., Tobias, S. M. and Weiss, N. O., 'For how long will the current grand maximum of solar activity persist?', *Geophysical Research Letters*, vol. 35, no. 20 (2008)

Babcock, H. W., 'The topology of the sun's magnetic field and the twenty two year cycle', *Astrophysical Journal*, vol. 133 (1961), 572

Bahcall, J. N., 'Solar models: An historical overview', *Nuclear Physics B Proceedings Supplements*, vol. 118 (2003), 77–86

Bahcall, J. N., Serenelli, A. M. and Basu, S., '10,000 standard solar models: A Monte Carlo simulation', *Astrophysical Journal Supplement Series*, vol. 165, no. 1 (2006), 400–431

Baliunas, S. L., Donahue, R. A., Soon, W. and Henry, G. W., 'Activity cycles in lower main sequence and POST main sequence stars: The HK Project', in *The Tenth Cambridge Workshop on Cool Stars, Stellar Systems and the Sun* (1998), p. 153

Baum, W. A., et al., 'Solar ultraviolet spectrum to 88 kilometres', *Physical Review*, vol. 70, no. 9–10 (1946), 781–2

Beck, J. G., 'A comparison of differential rotation measurements', *Solar Physics*, vol. 191, no. 1 (2000), 47–70

Berger, M. A. and Ruzmaikin, A., 'Rate of helicity production by solar rotation', *Journal of Geophysical Research*, vol. 105, no. A5 (2000), 10481–90

Biermann, L., 'Solar corpuscular radiation and the interplanetary gas', *The Observatory*, vol. 77 (1957), 109–10

Blanch, G., Lowan, A. N., Marshak, R. E. and Bethe, H. A., 'The internal temperature-density distribution of the Sun', *Astrophysical Journal*, vol. 94 (1941), 37

Brooks, D. H. and Warren, H. P., 'Establishing a connection between active region outflows and the solar wind: abundance measurements with EIS/Hinode', *Astrophysical Journal Letters*, vol. 727, no. 1 (2011), article id. L13

Brooks, D. H., Ugarte-Urra, I. and Warren, H. P., 'Full-Sun observations for identifying the source of the slow solar wind', *Nature Communications*, vol. 6 (2015)

Cahan, D., 'The awarding of the Copley Medal and the "discovery" of the law of conservation of energy: Joule, Mayer and Helmholtz revisited', *Notes Rec. R. Soc.*, vol. 66 (2012), 125–39

Carrington, R. C., 'Description of a singular appearance seen in the Sun on September 1, 1859', *Monthly Notices of the Royal Astronomical Society*, vol. 20 (1859), 13–15

Chaplin, W. C., et al., 'Observing the Sun with the Birmingham Solar-Oscillations Network (BISON)', *The Observatory*, vol. 116 (1996), 32–3

Chapman, A., 'Thomas Harriot: the first telescopic astronomer', *Journal of the British Astronomical Association*, vol. 118, no. 11 (2008), 315–25

Chapman, S. and Zirin, H., 'Notes on the solar corona and the terrestrial ionosphere', *Smithsonian Contribution to Astrophysics*, vol. 2 (1957), 1

Chubb, T. A., Friedman, H., Kreplin, R. W. and Kupperian, J. E., Jr, 'Lyman alpha & X-ray emissions during a small solar flare', *Journal of Geophysical Research*, vol. 62, no. 3 (1957), 389–98

Clarke, T. E., Kronberg, P. P. and Böhringer, H., 'A new radio-X-Ray probe of galaxy cluster magnetic fields', *Astrophysical Journal*, vol. 547, no. 2 (2001), L111–L114

Cliver, E. W. and Dietrich, W. F., 'The 1859 space weather event revisited: limits of extreme activity', *Journal of Space Weather and Space Climate*, vol. 3, no. 15 (2013)

Cowling, T. G., 'Convection in stars', *The Observatory*, vol. 58 (1935), 243–7

Crowther, P. A., et al., 'The R136 star cluster hosts several stars whose individual masses greatly exceed the accepted 150 M$_\odot$ solar stellar mass limit', *Monthly Notices of the Royal Astronomical Society*, vol. 408, no. 2 (2010), 731–51

Démoulin, P., et al., 'What is the source of the magnetic helicity shed by CMEs? The long-term helicity budget of AR 7978', *Astronomy and Astrophysics*, vol. 382 (2002), 650–65

Dicke, R. H., 'The Sun's rotation and relativity', *Nature*, vol. 202, no. 4931, 1964, 432–5

Dicke, R. H. and Goldenberg, H. M., 'Differential rotation and the solar oblateness', *Nature*, vol. 214, no. 5095 (1967), 1294–6

Dorling, E., 'Ariel I and the beginnings of British space science', *The Observatory*, vol. 113 (1993), 250–55

Dungey, J. W., 'Interplanetary magnetic field and the Auroral zones', *Physical Review Letters*, vol. 6, no. 2 (1961), 47–9

Duvall, T. L., Jr, 'Large-scale solar velocity fields', *Solar Physics*, vol. 63 (1979), 3–15

Eddington, A. S., 'The deflection of light during a solar eclipse', *Nature*, vol. 104 (1919), 372

—, 'The source of stellar energy', *Nature*, vol. 115, no. 2890 (1925), 419–20

Edlén, B., 'Die Deutung der Emissionslinien im Spektrum der Sonnenkorona. Mit 6 Abbildungen', *Zeitschrift für Astrophysik*, vol. 22 (1942), 30

English, R. A., Benson, R. E., Bailey, J. V. and Barnes, C. M., 'Apollo experience report – protection against radiation'

Fletcher, L. and Hudson, H. S., 'Impulsive phase flare energy transport by large-scale Alfvén waves and the electron acceleration problem', *Astrophysical Journal*, vol. 675, no. 2 (2008), 1645–55

Gough, D. O., et al., 'The seismic structure of the Sun', *Science*, vol. 272, no. 5266 (1996), 1296–1300

Green, L. M. and Kliem, B., 'Flux rope formation preceding coronal mass ejection onset' *The Astrophysical Journal Letters*, vol. 700 (2009), issue 2, L83–L87

Green, L. M., et al., 'The magnetic helicity budget of a CME-prolific active region', *Solar Physics*, vol. 208, no. 1 (2002), 43–68

Green, L. M., Kliem, B., Török, T., van Driel-Gesztelyi, L. and Attrill, G. D. R., 'Transient coronal sigmoids and rotating erupting flux ropes', *Solar Physics*, vol. 246, no. 2 (2007), 365–91

Greenstein, J. L., 'Biographical memoirs. Essay on Robert B. Leighton', *National Academy of Sciences Biographical Memoirs*, vol. 75 (1998)

Grotrian, W., 'Zur Frage der Deutung der Linien im Spektrum der Sonnenkorona', *Die Naturwissenschaften*, vol. 27, no. 13 (1939), 214

Haigh, J. D., Blackburn, M. and Day, R., 'The response of tropospheric circulation to perturbations in lower-stratospheric temperature', *Journal of Climate*, vol. 18, no. 17 (2005), 3672–85

Hale, G. E., 'On the probable existence of a magnetic field in sunspots', *Astrophysical Journal*, vol. 28 (1908), 315

—, '"Solar vortices". Contributions from the Mount Wilson solar observatory', *Carnegie Institution of Washington*, vol. 26, 1–17

—, 'Solar vortices and magnetic fields', *The Observatory*, vol. 32 (1909), 311–15

—, '"Solar vortices and magnetic fields". Contributions from the Mount Wilson solar observatory', *Astrophysical Journal*, vol. 28 (1908), 100

—, 'The Sun as a research laboratory', *Journal of the Franklin Institute*, vol. 204, no. 1 (1927), 19–28

Hardy, R., 'Theophrastus's observation of sunspots', *Journal of the British Astronomical Association*, vol. 101, no. 5 (1991), 261

Hathaway, D. H., et al., 'GONG observations of solar surface flows', *Science*, vol. 272, no. 5266 (1996), 1306–9

Hathaway, D. H. and Rightmire, L., 'Variations in the Sun's meridional flow over a solar cycle', *Science*, vol. 327, no. 5971 (2010), 1350ff

Herschel, J., 'Observations of Halley's comet with remarks on its physical condition and that of comets in general', in *Results of astronomical observations made during the years 1834, 5, 6, 7, 8 at the Cape of good hope*, 1847

Herschel, W., 'Observations tending to investigate the nature of the Sun, in order to find the causes or symptoms of its variable emission of light and heat; with remarks on the use that may possibly be drawn from solar observations', *Philosophical Transactions of the Royal Society of London,* vol. 91 (1801), 265–318

—, 'On the nature and construction of the Sun and fixed stars', *Philosophical Transactions of the Royal Society of London*, vol. 85, 46–72

Hilberg, R. H., 'Radiation protection for Apollo missions – case 340'

Hodgson, R., 'On a curious appearance seen in the Sun', *Monthly Notices of the Royal Astronomical Society*, vol. 20 (1859), 15–16

Howe, R., 'Solar interior rotation and its variation', *Living Reviews in Solar Physics,* vol. 6, no. 1 (2009)

Hu, S., Kim, M. Y., McClellan, G. and Cucinotta, F. A., 'Modeling the acute health effects of astronauts from exposure to large solar particle events', *Health Physics Society Journal*, vol. 96 (2009), 465–76

Johnson, M. J., 'Address delivered by the President, M. J. Johnson, Esq., on presenting the Gold Medal of the Society to M. Schwabe', *Monthly Notices of the Royal Astronomical Society*, vol. 17, 126, 185

Komm, R. W., Howard, R. F. and Harvey, J. W., 'Meridional flow of small photospheric magnetic features', *Solar Physics*, vol. 147, no. 2 (1993), 207–23

Kosovichev, A. G. and Zharkova, V. V., 'X-ray flare sparks quake inside Sun', *Nature*, vol. 393, no. 6683 (1998), 317–18

Labonte, B. J. and Howard, R., 'Solar rotation measurements at Mount Wilson. III – Meridional flow and limbshift', *Solar Physics*, vol. 80 (1982), 361–72

Lane, H. J., 'On the theoretical temperature of the Sun, under the hypothesis of a gaseous mass maintaining its volume by its internal heat, and depending on the laws of gases as known to terrestrial experiment', *American Journal of Science*, vol. 50 (1870), 57–74

Leibacher, J. W., Noyes, R. W., Toomre, J. and Ulrich, R. K, 'Helioseismology', *Scientific American*, vol. 253 (1985), 48–57

Leighton, Robert B., Noyes, Robert W. and Simon, George W., 'Velocity fields in the solar atmosphere. I. Preliminary report', *Astrophysical Journal*, vol. 135 (1962), 474

Leka, K. D., Canfield, R. C., McClymont, A. N. and van Driel-Gesztelyi, L., 'Evidence for current-carrying emerging flux', *Astrophysical Journal*, vol. 462 (1996), 547

Lockwood, M. and Fröhlich, C., 'Recent oppositely directed trends in solar climate forcings and the global mean surface air temperature', *Proceedings of the Royal Society of London A: Mathematical, Physical and Engineering Sciences*, vol. 463 (2007), 2447–60

Lockwood, M., Harrison, R. G., Woollings, T. and Solanki, S. K., 'Are cold winters in Europe associated with low solar activity?', *Environmental Research Letters*, vol. 5, no. 2 (2010)

Lockwood, M., Owens, M., Barnard, L., Davis, C. and Thomas, S., 'Solar cycle 24: What is the Sun up to?', in *Astronomy & Geophysics*, vol. 53, no. 3 (2012), 3.09–3.15

Lockyer, J. N., 'Spectroscopic observations of the Sun', *Proceedings of the Royal Society of London*, vol. 15 (1866), 256–8

Low, B. C., 'Solar activity and the corona', *Solar Physics*, vol. 167, no. 1–2 (1996), 217–65

Maehara, H., et al., 'Superflares on solar-type stars', *Nature*, vol. 485, no. 7399 (2012), 478–81

McComas, D. J., et al., 'Weaker solar wind from the polar coronal holes and the whole Sun', *Geophysical Research Letters*, vol. 35, no. 18 (2008)

Mitalas, R. and Sills, K. R., 'On the photon diffusion time scale for the sun', *Astrophysical Journal*, vol. 401, no. 2 (1992), 759–60

Mitchell, W. M., 'The history of the discovery of the solar spots', *Popular Astronomy*, vol. 24 (1916), 22ff

Moore, R. and Rabin, D., 'Sunspots', *Annual Review of Astronomy and Astrophysics*, vol. 23, 239–66

Muñoz-Jaramillo, A., Nandy, D., Martens, P. C. H., and Yeates, A. R., 'A double-ring algorithm for modeling solar active regions: unifying kinematic dynamo models and surface flux-transport simulations', *Astrophysical Journal Letters*, vol. 720, no. 1 (2010), L20–L25

Narin, F., *Starfish Prime Report*, 1962

Neidig, D. F. and Cliver, E. W., 'A catalogue of solar white-light flares, including their statistical properties and associated emissions, 1859–1982', *AFGL Technical Report*, 1983

Newton, I., 'A letter of Mr. Isaac Newton, professor of the mathematicks in the University of Cambridge; containing his new theory about light and colours: sent by the author to the publisher from Cambridge, February 6, 1671–72; in order to be communicated to the R. Society', *Phil. Trans. R. Soc. Lond.*, vol. 6 (1671), 3075–87

November, L. J., Toomre, J., Gebbie, K. B. and Simon, G. W., 'The detection of mesogranulation on the Sun', *Astrophysical Journal, Part 2 – Letters to the Editor*, vol. 245 (1981), L123–L126

Parker, E. N., 'Dynamics of the interplanetary gas and magnetic fields', *Astrophysical Journal*, vol. 128 (1958), 664

—, 'The formation of sunspots from the solar toroidal field', *Astrophysical Journal*, vol. 121 (1955), 491

—, 'A history of the solar wind concept', in Bleeker, J. A., Geiss, J. and M. C. E. Huber, *The Century of Space Science*, New York: Springer, 2001

—, 'Something stirs under the Sun', *Nature*, vol. 379 (1996), 209–10

Pettersen, B. R., 'A review of stellar flares and their characteristics', *Colloquium on Solar and Stellar Flares*, vol. 121, no. 1–2 (1989), 299–312

Ribes, J. C. and Nesme-Ribes, E., 'The solar sunspot cycle in the Maunder minimum AD 1645 to AD 1715', *Astronomy and Astrophysics*, vol. 276 (1993), 549

Russell, H. N., 'On the composition of the Sun's atmosphere', *Astrophysical Journal*, vol. 70 (1929), 11

—, 'The properties of matter as illustrated by the stars', *Publications of the Astronomical Society of the Pacific*, vol. 33, no. 196 (1921), 275

Saar, S. H., 'Recent magnetic fields measurements of stellar magnetic fields', in *Stellar Surface Structure: Proceedings of the 176th Symposium of the International Astronomical Union, Held in Vienna, Austria, October 9–13, 1995*, eds. Klaus G. Strassmeier and Jeffrey L. Linsky, International Astronomical Union, Symposium no. 176, Kluwer Academic Publishers, Dordrecht, 1996, p. 237

Sadlier, G., Flytkær, R., Halterbeck M., and Pearce, W., 'Executive summary: the size & health of the UK space industry'

Sagan, C. and Mullen, G., 'Earth and Mars: evolution of atmospheres and surface temperatures', *Science*, vol. 177, no. 4043 (1972), 52–6

Scheiner, C., 'Rosa ursina sive sol', in *Andreas Phaeus, Braccuabi*, 1630

Shapiro, A., 'Bath-tub vortex', *Nature*, vol. 196 (1962), 1080–81

Shibata, K., et al., 'Can superflares appear on our sun?', *Publications of the Astronomical Society of Japan*, vol. 65, no. 3 (2013), article no. 49, 8

Solanki, S. K., et al., 'Unusual activity of the Sun during recent decades compared to the previous 11,000 years', *Nature*, vol. 431, no. 7012 (2004), 1084–7

Spiegel, E. A. and Zahn, J. P., 'The solar tachocline', *Astronomy and Astrophysics*, vol. 265, no. 1 (1992), 106–14

St Cyr, O. C., et al., 'Space Weather Diamond: a four spacecraft monitoring system', *Journal of Atmospheric and Solar-Terrestrial Physics*, vol. 62, no. 14 (2000), 1251–5

Steinhilber, F., Abreu, J. A., and Beer, J., 'Solar modulation during the Holocene', *Astrophysics and Space Sciences Transactions*, vol. 4, no. 1 (2008), 1–6

Stix, M., 'On the time scale of energy transport in the sun', *Solar Physics*, vol. 212, no. 1 (2003), 3–6

Strömgren, B., 'The boundary-value problem of the theory of stellar absorption lines', *Astrophysical Journal*, vol. 86 (1937), 1

Tobias, C. A., et al., 'Visual phenomena induced by cosmic rays and accelerated particles' (1972)

Tousey, R., 'Some results of twenty years of extreme ultraviolet solar research', *Astrophysical Journal*, vol. 149 (1967), 239

—, 'Survey of new solar results', *New techniques in Space Astronomy*, International Astronomical Union, Symposium no. 41, Dordrecht, Reidel, 1971, p. 233

Trefethen, L. M., et al., 'The bath-tub vortex in the Southern hemisphere', *Nature,* vol. 207 (1965), 1084–5

Usoskin, I. G., 'A history of solar activity over millennia', *Living Reviews in Solar Physics*, vol. 5, no. 3 (2013)

van Ballegooijen, A. A. and Martens, P. C. H., 'Formation and eruption of solar prominences', *Astrophysical Journal*, vol. 343 (1989), 971

van Driel-Gesztelyi, L., et al., 'Magnetic topology of active regions and coronal holes: implications for coronal outflows and the solar wind', *Solar Physics*, vol. 281, no. 1 (2012), 237–62

Walker, C. V., 'On magnetic storms and earth-currents', *Phil. Trans. R. Soc. Lond.*, vol. 151 (1861), 89–131

Wildt, R., 'Negative ions of hydrogen and the opacity of stellar atmospheres', *Astrophysical Journal*, vol. 90 (1939), 611

Wilson, A. and Maskelyne, N., 'Observations on the solar spots', *Philosophical Transactions*, vol. 64 (1683–1775), 1–30

Wittmann, A. D. and Xu, Z. T., 'A catalogue of sunspot observations from 165 BC to AD 1684', *Astronomy and Astrophysics Supplement Series*, vol. 70, no. 1 (1987), 83–94

Wollaston, W. H., 'A method of examining refractive and dispersive powers, by prismatic reflection', *Phil. Trans. R. Soc. Lond.*, vol. 92 (1802), 365–80

Zeeman, P., 'On the influence of magnetism on the nature of the light emitted by a substance', *Astrophysical Journal*, vol. 5 (1897), 332

Zhitnik, I. A., et al., 'Observations of the Sun and its spectrum at 9.5–200 A', *Cosmic Research*, vol. 5, 237

Books

Eddington, A. S., *The Internal Constitution of the Stars*, Cambridge University Press, 1926

Foukal, P. V., *Solar Astrophysics*, Wiley-VCH, 2013

Hale, G. E. and Nicholson, S. B., *Magnetic Observations of Sunspots*, Carnegie Institution of Washington, 1938

Howard, T., *Coronal Mass Ejections: An Introduction*, Springer, 2011

Massey, H. S. W. and Robins, M. O., *History of British Space Science*, Cambridge University Press, 1986

Payne, C. H., *Stellar Atmospheres: A Contribution to the Observational Study of High Temperature in the Reversing Layers of Stars*, Harvard Observatory, 1925

Phillips, K. J. H., *Guide to the Sun*, Cambridge University Press, 1992

Schrijver, C. J. and Zwaan, C., *Solar and Stellar Magnetic Activity*, Cambridge University Press, 2008

Stix, M., *The Sun: An Introduction*, Springer, 2002

Tayler, R. J., *The Sun as a Star*, Cambridge University Press, 1996

Acknowledgements

Right from the start I had support from Sarah Green, Ben Dixon, Valerie Green, Alan Green, Julia Green, Simon Green, Anna Wu and Mike Wu, who all gave great advice on how to turn an academic subject into a human story. And my husband, Matt Parker, gets special thanks for the constant supply of tea and patience.

I would like to thank the following colleagues who have read and commented on sections of the book at various points along the way: Mitch Berger, Dave Brooks, Paul Cannon, Bill Chaplin, Paul Crowther, Len Culhane, Pascal Démoulin, Lidia van Driel-Gesztelyi, Peter Gallagher, Joanna Haigh, Hugh Hudson, Bernhard Kliem, Mike Lockwood, Andrew Richards, Sami Solanki, Kinwah Wu and my enthusiastic Ph.D. student Stephanie Yardley. Len Culhane, who was my Ph.D. supervisor, has always been kind enough to answer all my questions about the origin of the UK and European space programmes as well as the history of the Mullard Space Science Laboratory, sharing his recollections of the rapid pace of discovery and international collaborations. And I would like to thank George Doschek, who shared with me his stories about the research carried out with Skylab. When it comes to space weather, Doug Biesecker has for many years been a source of inspiration and has shared with me the history of space weather monitoring in the US. Then there is Lewis Dartnell, who gave me insight into the effects of radiation on the human body. And much of the research around the Thames freezing over in 1814 was done with the team I worked with on the *Killer Storms and Cruel Winters* programme for BBC4.

So many of the stories on these pages have been shared amongst members of the solar physics community, both in the UK and abroad.

And the stories I didn't have space to tell would fill many more books. It was painful having to leave so many important and interesting ones out. But I hope I have captured some old favourites here, and also some new stories too. I owe a lot to the solar and astrophysics community, who have always been keen to discuss the science and the history of our research area. In particular, my group at UCL gets a special mention, including past members Bernhard Kliem and Tibor Török, who taught me so much about the physics of magnetic fields and numerical modelling of coronal mass ejections. Then, there are three very special people whom I caught the solar physics bug from in the first place through their infectious enthusiasm, curiosity and support: Lidia van Driel-Gesztelyi, Pascal Démoulin and Cristina Mandrini. Thank you.

The wider community also includes the instrument teams who built such successful telescopes and who are now building the solar satellites of the future. They work behind the scenes but deserve huge recognition for their innovation and ability to overcome the challenges presented by placing telescopes in space.

The process of writing this book, despite knowing the subject area, was lengthy. So I want to say a huge thank you to my two editors: Will Hammond, who got me off the blocks, and Daniel Crewe, who saw me across the finishing line. Together they taught me how to take facts and figures and weave them into a narrative that is colourful and meaningful.

Various people helped with the images that are a necessary part of such a visual subject, including Sian Prosser at the Royal Astronomical Society and John Grula, Cindy Hunt and Dan Kohne of the Carnegie Observatories. I also would like to thank the Leverhulme Trust and the Royal Society, which have supported my research through Fellowship schemes during this time.

Index